AROMATASE INHIBITORS: TYPES, MODE OF ACTION AND INDICATIONS

Cancer Etiology, Diagnosis and Treatments Series

Cell Apoptosis and Cancer
Albina W. Taylor
2007 ISBN: 1-60021-506-8

Chronic Lymphocytic Leukemia Research Focus
Chadi Nabhan
2007ISBN: 1-60021-526-2

Cervical Cancer Research Trends
Eleanor P. Bankes
2007 ISBN: 1-60021-648-x

Lung Cancer in Women
Varetta N. Torres
2008 ISBN: 1-60021-659-5

Cancer Research at the Leading Edge
Ignatius K. Martakis
2008 ISBN: 1-60021-728-1

Chronic Lymphocytic Leukemia: New Research
Inès B. Moreau
2008 ISBN: 978-1-60456-081-7

Cancer and Stem Cells
Thomas Dittmar and Kurt S. Zander
2008 ISBN: 978-1-60456-478-5

Cancer Prevention Research Trends
Louis Braun and Maximilian Lange
2008 ISBN: 978-1-60456-639-0

Clinical, Genetic and Molecular Precursor Features in Colorectal Neoplasia
Kjetil Søreide and Håvard Søiland
2008 ISBN: 978-1-60456-714-4

Human Polyomaviruses: Molecular Mechanisms for Transformation and their Association with Cancers
Ugo Moens, Marijke Van Gheule and Mona Johannessen
2009 ISBN: 978-1-60692-812-7

Anticancer Drugs: Design, Delivery and Pharmacology
Peter Spencer and Walter Holt
2009 ISBN: 978-1-60741-004-1

Molecular Therapy of Breast Cancer: Classicism Meets Modernity
Marc Lacroix
2009 ISBN: 978-1-60741-593-0

Aromatase Inhibitors: Types, Mode of Action and Indications
Jean R. Lamonte
2009 ISBN: 978-1-60741-711-8

AROMATASE INHIBITORS: TYPES, MODE OF ACTION AND INDICATIONS

Jean R. Lamonte
EDITOR

Nova Biomedical Books
New York

For permission to use material from this book please contact us:
Telephone 631-231-7269; Fax 631-231-8175
Web Site: http://www.novapublishers.com

Library of Congress Cataloging-in-Publication Data

Available upon request

ISBN: 978-1-60741-711-8

Published by Nova Science Publishers, Inc. ✢ New York

Contents

Preface		**vii**
Chapter I	Anti-Estrogen Agents; Aromatase Inhibitors: Past, Present and Future *Masanori Shimodaira and Tomohiro Nakayama*	**1**
Chapter II	Aromatase Inhibitors: Types, Mode of Action and Indications *Helena Aguilar and Ander Urruticoechea*	**21**
Chapter III	Aromatase Inhibitors *Mosammat Rashida Begum and Maruf Siddiqui*	**39**
Chapter IV	Upfront vs. Sequential Aromatase Inhibitors: An Ongoing or Settled Debate? *Tallal Younis and Chris Skedgel*	**55**
Chapter V	Aromatase Expression in the Endometrium and its Role in the Development of Uterine Pathology *Hugo Maia Jr, Julio Casoy, Jorge Valente Filho and Elsimar Coutinho*	**75**
Chapter VI	Aromatase Inhibitors for the Medical Treatment of Endometriosis *Kaei Nasu, Akitoshi Tsuno, Akitoshi Yuge, Masakazu Nishida and Hisashi Narahara*	**95**
Chapter VII	An Update of the Use of Aromatase Inhibitors in the Management of Breast Cancer *S. Tang, D. A. L. Morgan and K. L. Cheung*	**113**
Chapter VIII	Aromatase Inhibitors in Ovulation Induction. Myth or Reality? *Nikolaos P. Polyzos, Spyridon Tzioras, Davide Mauri, Maria Tsappi, Antonis Valachis, Charalambos Dritsas, Kamposioras Konstantinos, Charalambos Panou and Lamprini Tsali*	**147**

Chapter IX The Use of Aromatase Inhibitors in Ovarian Cancer and Identification of Responsive Tumors **163**
S. P. Langdon, G. Walker, D. Faratian, C. Gourley, A. R. W. Williams, I. Um, J. F. Smyth and D. J. Harrison

Index **175**

Preface

Aromatase inhibitors (AI) are a class of drugs used in the treatment of breast cancer and ovarian cancer in postmenopausal women. Aromatase is an enzyme that synthesizes estrogen and aromatase inhibitors block the synthesis of estrogen, which is needed in some cancers to grow. As breast tissue is stimulated by estrogens, decreasing their production is a way of suppressing recurrence of the breast tumor tissue. In contrast to pre-menopausal women, in whom most of the estrogen is produced in the ovaries, in post-menopausal women most of the body's estrogen is produced in the adrenal gland from the conversion of androgens. Investigations are ongoing to look for other applications and researchers continue to study aromatase inhibitors as a way to stimulate ovulation or suppress estrogen production in endometriosis. This new book presents the latest research from around the world in this dynamic field.

Chapter I - Aromatase is an enzyme that catalyzes the conversion of androgens to estrogens, the final critical step of estrogen synthesis. It is present in various tissues, such as ovaries, placenta, brain and adipose tissue. Pathologically, abnormal overexpression of aromatase plays an important role in the development of breast cancer; thus, aromatase inhibitors (AIs) are commonly used as hormone therapy for postmenopausal patients with estrogen sensitive breast cancer. According to the order of their clinical development, AIs are described as first-, second- or third-generation inhibitors. In the late 1980s, research focused on developing third-generation AI agents with increased potency and higher selectivity for estrogen synthesis. The identification of two different mechanisms for AIs led to the development of two types of third-generation AIs. Type I AIs, like exemestane, are known as steroidal inhibitors and have androstenedione-like structures, binding to androgen receptors with low affinity. Type II AIs, such as anastrozole and letrozole, are known as non-steroidal inhibitors and are non-competitive inhibitors of aromatase. At present, these are widely used as the first-line drugs in the endocrine treatment of hormone-dependent breast cancer because of their effective activity and general tolerability when compared with tamoxifen. Although third-generation AIs have demonstrated superior efficacy and better overall safety compared with tamoxifen, postmenopausal women taking AIs suffered more bone loss than women taking tamoxifen. Patients require regular bone mineral density monitoring, and bisphosphonates are being evaluated to determine whether they may protect long-term bone health. In addition to AIs, natural anti-aromatase chemicals such as flavones and lignans have been identified in plants. Efforts to develop new lines of AIs obtained from these phytochemicals have been initiated in several laboratories. There has also been significant progress in the understanding of the structure-function relationship of aromatase. Such information helps to evaluate the reaction mechanisms of aromatase, to examine the binding characteristics of AIs, and to explain the molecular basis for the low catalytic activity of the natural variant.

Chapter II - Tamoxifen, a selective estrogen receptor modulator (SERM), has been used for many decades as the "gold standard" endocrine treatment for patients with hormone-receptor-positive breast cancer. In the last few years, third generation aromatase inhibitors (AIs), either steroidal (exemestane) or non-steroidal (anastrozole, letrozole), have shown to be an effective alternative to tamoxifen in postmenopausal patients. These agents act blocking the aromatase enzyme which converts androgens into estrogens. They are associated with superior activity and better general tolerability when compared with tamoxifen both in the adjuvant and metastatic settings.

The potencial differences between steroidal aromatase inactivators and the non-steroidal aromatase inhibitors, manifested by the lack of total cross-resistance between compounds, underline the importance of exploiring the consequences of structural differences between aromatase inhibitors and inactivators.

A major obstacle to the clinical use of endocrine therapy is the occurrence of primary or acquired endocrine resistance. A better understanding of the multidirectional cross-talk between growth factor pathways, the estrogen receptors (both at the genomic and non genomic level), and estrogen receptor cofactors has shown to be fundamental to overcome or prevent resistance to endocrine therapy.

Thus, future research exploring the mechanisms of action and resistance to aromatase inhibitors will lead to the development of more effective therapeutic strategies.

Chapter III - Aromatase inhibitors (AIs) are anti-estrogenic agents, which exert their action by inhibiting aromatase enzyme. AIs are a newer class of drugs that inhibit the function of the aromatase enzyme. In postmenopausal women, AIs can suppress most of the peripheral aromatase activity, leading to profound estrogen deprivation. AIs are classified as first, second or third generation according to their sequential development. Aromatization is the final step in steroid biosynthesis and, therefore, aromatase is a suitable target for selective inhibition. Aromatase is expressed primarily in the ovary and also in central and peripheral tissues, fat, muscle, liver, and breast. Normal and malignant breast tissue contributes to the peripheral synthesis of estrogens. Studies have consistently shown that lifetime exposure to estrogens increase the risk of breast cancer. The degree of risk is increased by persistently elevated blood concentrations of estrogen. AI can effectively reduce the estrogen level hence improves breast cancer including advanced stage.

Aromatase enzyme is demonstrated locally in endometriotic implants, which indicates local production of estrogen by these implants. Thus, extraovarian aromatse is likely to cause a resistance to GnRHa treatment via persistence of local and peripheral estrogen production during such treatment. In this situation aromatase inhibitors are effective to reduce the synthesis of both ovarian and extraovarian estrogen.

Anti-estrogens have lot of potential to be used as oral ovulation inducing agent. In the late 1990s, aromatase inhibitors, including letrozole, began to be used to induce ovulation by being administered in the early part of the menstrual cycle. Estrogen production from all sources is blocked by inhibiting aromatization, releasing the hypothalamic-pituitary axis from estrogenic negative feedback and resulting in increased gonadotropin secretion and ovarian follicular stimulation.

There may be potential new applications for aromatase inhibitors including improving implantation and prevention of ovarian hyperstimulation syndrome during assisted reproductive technology by reducing supraphysiologic level of estrogen.

Regarding safety concern different studies showed that there were no increased rates of major and minor malformations beyond what would be expected in the general population (1-3%).

It is not far away that aromatase inhibitors will possibly emerge as first line ovulation inducing drug for anovulatory or ovulatory infertility following further large-scale, multicentric, randomized controlled research and clinical trials.

Chapter IV - Adjuvant hormonal therapy with an aromatase inhibitor has become a standard treatment option for post menopaual women with breast cancer. Compared to tamoxifen alone, upfront aromatase inhibitor and sequential tamoxifen - aromatase inhibitor strategies have been shown to improve the recurrence risk of breast cancer in a number of clinical trials. Despite the higher costs associated with aromatase inhibitors, compared to tamoxfien, both approaches have also been shown to be cost effective strategies with good value for money. As such, aromatase inhibitors are currently indicated as adjuvant hormonal therapy in post menopausal women with breast cancer considering their favorable clinical and economic profile. This chapter reviews the current evidence, from a clinical and an economic perspective, with regard to the preferred adjuvant strategy of aromatase inhibitor whether upfront or sequential treatment. The evolution of clinical evidence from indirect across studies analyses / modeling to the first results of a direct randomized study comparing both approaches is discussed. Economic evaluations based on indirect comparison across clinical studies are reviewed, and a direct head to head comparison based on BIG 1-98 study is presented.

Chapter V - Aromatase expression in the eutopic endometrium is triggered by constant exposure to inflammatory mediators such as prostaglandins, which are produced during menstruation and in the proliferative phase of the menstrual cycle. The presence of aromatase in the endometrial cells is pivotal for the development of endometriosis in the peritoneal cavity, since local estrogen production may be the mechanism that suppresses the phagocytosis of these cells by macrophages and NK cells. The severity of endometriosis correlates positively with the intensity of aromatase expression in the eutopic endometrium, thus corroborating the hypothesis that the progression of endometriosis depends on the constant shedding of aromatase-positive cells through retrograde menstruation. The coexistence of other pathologies with endometriosis is also associated with the presence of aromatase expression in the endometrium. This has been observed in cases of submucous/intramural myomas, endometrial polyps and adenomyosis. In all these cases, the presence of aromatase was associated with increased inflammation in the endometrium as determined by the presence of NF-kappa.B in cell nuclei. The number of NF-kappa.B-positive nuclei was significantly greater in the endometrium of pathology-harboring uteri compared to normal controls. This was observed both during the proliferative phase and prior to the onset of menstruation, thus suggesting a proinflammatory role for estrogens and an antiinflammatory effect of progesterone in the endometrium. Endogenous progesterone or progestins such as gestodene or levonorgestrel when used orally or delivered locally in the uterine cavity are potent inhibitors of aromatase expression in the endometrium and this effect probably involves gene-silencing through transcription inhibition. The decrease in aromatase expression induced by progestins was accompanied by a reduction in the expression of enzymes such as cyclooxygenase type 2 (Cox-2) or angiogenic factors such as VEGF in the endometrium. Inflammation in the endometrium was also diminished by progesterone or progestins and this mechanism involved the inhibition of NF-kappa.B activation. These findings support the role of progestins as anti-aromatase and antiinflammatory agents in the current management of endometriosis and other gynecological pathologies. In fact, the continuous use of combined oral contraceptives containing gestodene or the use of levonorgestrel-releasing intrauterine systems were found to be effective in preventing not only the recurrence of endometriosis but also myoma-related menorrhagia, thereby supporting the causal relationship between aberrant aromatase expression in the endometrium, inflammation and the development of pathology.

Chapter VI - Endometriosis, a common, benign, estrogen-dependent disease affecting 3-10% of women of reproductive age, is characterized by the ectopic growth of endometrial tissue found primarily in the peritoneum, ovaries, and rectovaginal septum. Women suffering from endometriosis may present with chronic pelvic pain, dysmenorrhea, dyspareunia, and/or subfertility. The prevalence of endometriosis in women with pelvic pain and/or subfertility is estimated to be between 20-90%; thus, it is one of the most frequently encountered benign gynecological problems.

Various surgical and medical strategies have been used to treat endometriosis. Medical therapies historically have included contraceptive steroids, progestogens, and agonists of gonadotropin-releasing hormone, as well as androgens and non-steroidal anti-inflammatory agents. Current medical treatment aims to inhibit the growth of endometriotic implants by suppressing ovarian steroids and inducing a hypoestrogenic state. However, treatments for endometriosis that aim to lower circulating estradiol concentrations can be used for only a limited time due to unacceptable side effects. In addition, high recurrence rates of up to 45% after the completion of medical treatments remain a significant problem. Therefore, novel therapeutic strategies are necessary to improve the clinical management of patients with endometriosis.

Aromatase P450 is a key enzyme for estrogen biosynthesis, because aromatase catalyzes the conversion of androstenedione and testosterone to estrone and estradiol. Increasing evidence suggests that endometrial aromatase expression is involved in the pathogenesis of endometriosis. Recent studies show that aromatase expression is consistently found in endometriotic lesions and in the eutopic endometrium from patients with endometriosis, whereas it is absent in eutopic endometrium from women without endometriosis. The aromatase enzyme, which is demonstrated to be present locally in endometriotic implants, is indicative of estrogen biosynthesis in these lesions. Local estrogen production by these implants could contribute to the progression of endometriosis even during treatment with GnRH agonists, which inhibit only ovarian production of estrogen. A non-steroidal aromatase inhibitor is a highly potent and specific suppressor of estrogen production locally and systemically. Several reports have demonstrated the therapeutic role of the aromatase inhibitor in patients with endometriosis.

This paper is a review of the recent literature concerning basic and clinical research on the use of aromatase inhibitors to medically treat endometriosis. The mechanisms of action, efficacy, and adverse effects of aromatase inhibitors in the treatment of endometriosis are discussed.

Chapter VII - With the increasing use of endocrine therapy in the treatment of breast cancer, the role of aromatase inhibitors in the management of hormone receptor positive postmenopausal breast cancer has been thoroughly investigated. Third generation aromatase inhibitors (anastrozole, letrozole and exemestane) are very well tolerated, with potentially manageable side effects. They have an established role in the treatment of advanced disease. They have also been shown to be more effective than tamoxifen when used as adjuvant treatment in early primary breast cancer, either when used upfront, as sequential / switch therapy after two to three years of tamoxifen or as extended therapy after five years of tamoxifen. Recent studies of their use in premenopausal women (alongside ovarian suppression therapy) and as neoadjuvant treatment have shown promising results. Their potential roles in chemoprevention and as combination therapies (with biological therapy or a pure anti-oestrogen) are being investigated in ongoing trials. Further studies need to be conducted to identify tumour and patient prognostic and predictive factors that may improve patient selection for optimal targeted therapy with aromatase inhibitors.

Chapter VIII - Third generation aromatase inhibitors, drugs of choice for the treatment of hormone responsive breast cancer, have been introduced in clinical trials as potential agents for ovulation induction. Their efficacy has been tested against traditional ovulation induction regimens in women with anovulatory and other causes of infertility. However, while early data clearly supported the superiority of letrozole and anastrozole in ovulation induction, more recent publications' results were modest. In an attempt to clarify the discrepancies among published trials for aromatase inhibitors in ovulation induction the authors methodologically reviewed the available evidence and recorded quality characteristics. The authors separately assessed the level of evidence provided; sample size, statistical design, primary outcomes, secondary outcomes, side-effects and fetal outcomes in letrozole and anastrozole trials. Whereas an adequate number of Level I evidence was available, the statistical design and quality parameters were poorly reported. Primary outcomes (pregnancy rates) were not consistent among trials and data regarding secondary outcomes, side-effects and fetal outcomes were systematically omitted. Eventually, it is still unclear whether these drugs may represent a new treatment modality for ovulation induction. Data is conflicting; trials are small in sample size and suffer from quality and statistical design biases. A large multicenter and properly designed randomized trial is required if a solid conclusion is to be made on the role of letrozole in ovulatory induction.

Chapter IX - It has long been recognised that many ovarian cancers express estrogen receptors and their growth may be sensitive to anti-estrogen strategies. Aromatase is present in normal ovaries and in between 33 and 81% of ovarian cancers; its inhibition will result in a reduction of both circulating and local levels of estrogen. Recently, several aromatase inhibitors have been investigated in phase II clinical trials of recurrent ovarian cancer. Five trials of letrozole have now been completed in ovarian cancer with consistent and encouraging results. A significant percentage of patients derive some clinical benefit as indicated by change in serum levels of CA125, the surrogate biomarker of ovarian cancer tumor load. Between 10 and 36 % of patients have a CA125 response while a further 18 to 36% patients achieve stabilisation of CA125. The use of ovarian cancer cell line models has helped identify panels of estrogen-regulated genes whose expression can be linked with response. Several studies have now demonstrated the feasibility of identifying a sensitive patient subgroup by measurement of these biomarkers within primary cancers and their semi-quantitative expression can be linked with the surrogate marker of disease response and stabilisation after letrozole treatment. This review summarises the clinical studies undertaken to date in ovarian cancer and describes the progress made towards predicting a cohort who will respond to treatment.

In: Aromatase Inhibitors: Types, Mode of Action and Indications ISBN: 978-1-60741-711-8
Editor: Jean R. Lamonte

Chapter I

Anti-Estrogen Agents; Aromatase Inhibitors: Past, Present and Future

Masanori Shimodaira[a,b] and Tomohiro Nakayama[a,c,*]

[a]Division of Laboratory Medicine, Department of Pathology and Microbiology, Nihon University School of Medicine, Tokyo, Japan

[b]Divisions of Hematology, Endocrinology and Metabolism, Tokyo Metropolitan Hiroo Hospital, Tokyo, Japan

[c]Division of Nephrology and Endocrinology, Department of Medicine, Nihon University School of Medicine, Tokyo, Japan

Abstract

Aromatase is an enzyme that catalyzes the conversion of androgens to estrogens, the final critical step of estrogen synthesis. It is present in various tissues, such as ovaries, placenta, brain and adipose tissue. Pathologically, abnormal overexpression of aromatase plays an important role in the development of breast cancer; thus, aromatase inhibitors (AIs) are commonly used as hormone therapy for postmenopausal patients with estrogen sensitive breast cancer. According to the order of their clinical development, AIs are described as first-, second- or third-generation inhibitors. In the late 1980s, research focused on developing third-generation AI agents with increased potency and higher selectivity for estrogen synthesis. The identification of two different mechanisms for AIs led to the development of two types of third-generation AIs. Type I AIs, like exemestane, are known as steroidal inhibitors and have androstenedione-like structures, binding to

[*] Corresponding author: Tomohiro Nakayama, M.D. Division of Laboratory Medicine, Department of Pathology and Microbiology, Nihon University School of Medicine, Nihon University School of Medicine. 30-1 Ooyaguchi-kamimachi, Itabashi-ku, Tokyo 173-8610, Japan. Tel: +81-3-3972-8111 (Ext. 8205); Fax: +81-3-5375-8076. E-mail: tnakayam@med.nihon-u.ac.jp

androgen receptors with low affinity. Type II AIs, such as anastrozole and letrozole, are known as non-steroidal inhibitors and are non-competitive inhibitors of aromatase. At present, these are widely used as the first-line drugs in the endocrine treatment of hormone-dependent breast cancer because of their effective activity and general tolerability when compared with tamoxifen. Although third-generation AIs have demonstrated superior efficacy and better overall safety compared with tamoxifen, postmenopausal women taking AIs suffered more bone loss than women taking tamoxifen. Patients require regular bone mineral density monitoring, and bisphosphonates are being evaluated to determine whether they may protect long-term bone health. In addition to AIs, natural anti-aromatase chemicals such as flavones and lignans have been identified in plants. Efforts to develop new lines of AIs obtained from these phytochemicals have been initiated in several laboratories. There has also been significant progress in the understanding of the structure-function relationship of aromatase. Such information helps to evaluate the reaction mechanisms of aromatase, to examine the binding characteristics of AIs, and to explain the molecular basis for the low catalytic activity of the natural variant.

Introduction

The essential role of estrogen in the female reproductive function has long been known, but recent evidence has suggested that estrogens also exert a wide range of biological effects on non-reproductive systems and organs [1]. Paradoxically, the role of estrogens in male reproduction has been clarified in the last 10 years [2], and this has affected some classical standpoints in endocrinology and diseases.Circulating estrogens are derived from the aromatization of circulating androgens by means of the enzymatic complex known as aromatase, which is responsible for the aromatization of the androgen A-ring. Aromatase catalyzes three consecutive hydroxylation reactions to convert C-19 androgen to C-18 estrogen, which is the final critical step of estrogen synthesis [4]. Aromatase is widely expressed in a number of tissues throughout the body: testes (Sertoli and Leydig cells), ovaries (granulosa cells and luteal corpus), breasts, adipose tissue, muscles, liver, brain (including hypothalamus), hair follicles and fibroblasts [5-9]. Estrogen acts through specific receptors belonging to the nuclear receptors superfamily [10]. Two different types of estrogen receptor (ER) have been characterized to date; ER-alpha (ER-α) and another subtype known as ER-beta (ER-β) [11]. Both are intranuclear receptors that, once activated, modulate transcription at the genomic level. Furthermore a non-genomic pathway for estrogen action has been demonstrated, probably involving a plasma-membrane interaction of the ER [12, 13].

Models of Estrogen Deficiency

In recent years, unique mice models lacking estrogen activity, together with the discovery of naturally occurring mutations leading to congenital estrogen deficiency in humans, have permitted better clarification of several physiological roles of estrogens.

Knockout mice [14] for ER-α for ER-β, and for both ERs have provided useful information on the activity of ERs in rodents. On the other hand, knockout mice for the aromatase gene are useful for studying the lack of circulating estrogen.

Estrogens were erroneously considered to be a prerequisite for human survival and to be necessary for blastocyst implantation [15]. Thus, aromatase deficiency was regarded as incompatible with life, until the discovery of the first cases of human congenital estrogen deficiency [16, 17], which confirmed that aromatase deficiency is not a lethal defect and that estrogens are not necessary for embryonic or fetal development. Nevertheless, congenital estrogen deficiency in humans may be a unique model to study the effects of prenatal lack of estrogens on various organs and tissues, and the discovery of human mutations in both ER-α [18] and the aromatase gene [19] has strongly reinforced the idea of a key role for estrogens on male reproductive function. Human congenital estrogen deficiency is due to an inactivating mutation of the gene coding for the aromatase enzyme complex, which leads to a lack of estrogen synthesis, or to a mutation of the genes coding for estrogen receptors with a consequent lack of signal transduction, even in the presence of circulating estrogens. Aromatase-deficient females show hyperandrogenism and virilization at birth with ambiguous genitalia. In childhood, dysfunction in the LHRH-LH/FSH axis and progressive delay in bone age are seen. At puberty, they show primary amenorrhea, no breast development, exacerbation of virilization and the absence of growth spurt. The clinical phenotype in affected male subjects comprises tall stature, persistent linear growth and delayed bone age, osteopenia/osteoporosis, eunuchoid body proportions, differences in glucose-insulin balance and fertility impairment [20].

Breast Cancer, Estrogen and Estrogen Inhibitor (Tamoxifen)

Breast cancer is the most common cause of cancer death in women worldwide. Rates vary about five-fold around the world, but they are increasing, even in regions that until recently had low rates of disease [21] It is known that estrogens are closely involved in the growth of human breast carcinomas, and that the great majority of breast carcinomas express estrogen receptors. Recent studies have demonstrated that estrogens are locally produced and act on breast carcinoma tissue. Among these pathways, aromatase is a key enzyme for intratumoral production of estrogens in breast carcinomas, and aromatase inhibitors (AIs) are currently used in breast carcinoma in postmenopausal women as an estrogen deprivation therapy.

Tamoxifen is an oral anti-estrogen agent, first used in metastatic breast cancer in the early 1970s. It became the standard adjuvant therapy for women with early breast cancer following the first demonstration of efficacy more than 20 years ago [22]. Nevertheless, numerous limitations of tamoxifen have emerged with widespread use. In the landmark National Surgical Adjuvant Breast and Bowel Project B-14 trial, 66% of tamoxifen-treated patients experienced side effects when compared with 58% of patients given placebo [23]. Long-term adverse effects associated with 5 years of adjuvant tamoxifen include venous thromboembolic events, vaginal bleeding, vaginal discharge, ischemic cerebrovascular events, endometrial and uterine cancer, and hysterectomy Severe, potentially life-threatening events such as thrombosis were more likely to occur in patients aged >60 years. [24], and experiencing side effects significantly increases the likelihood of patients discontinuing tamoxifen therapy [25]. Recently, resistance to tamoxifen may develop [26], and therapy beyond 5 years is not recommended because neither further disease-free survival nor survival benefits are gained [27].

Large clinical trials were initiated in the late 1970s and early 1980s in order to test the drug's role as an adjuvant therapy in early-stage breast cancer. Observation of marked decreases in the development of contralateral breast cancer among tamoxifen recipients has suggested the potential for the drug in chemoprevention of breast cancer. The most recent analysis of the Early Breast Cancer Trialists' Collaborative Group (EBCTCG), which included information on 37,000 women in 55 trials of adjuvant tamoxifen, published in 2005, confirms the value of this oral anti-estrogen. Among women with ER-positive disease, the reduction in the recurrence rate and in the breast cancer death rate are significantly better either after 1–2 years or 5 years of tamoxifen treatment, with greater effects seen in the latter group. This indirect evidence that 1–2 years is less effective than 5 years of tamoxifen in ER-positive disease is highly significant with regard to recurrence, and for breast cancer mortality, and is supported by direct, randomised comparisons of different tamoxifen durations. In patients with ER-negative disease, tamoxifen may or not produce some benefit either after 1–2 years or 5 years of treatment. For women with tumors of unknown ER status, the benefits are slightly weaker than those for women with ER-positive disease [28].

With regard to its chemopreventive properties, a number of prevention trials with tamoxifen have shown a 48% reduction in ER-positive cancers [29]. Based on these findings, the United States Preventive Services Task Force recommended women with risk factors for breast cancer should discuss the potential benefit and harm of tamoxifen with their doctors [30]. The use of adjuvant tamoxifen has been associated with certain toxic effects. Most important, the development of endometrial cancer occurs at a rate 2 to 7 times greater, compared to untreated women [24, 31]. Tamoxifen is also associated with an increased incidence of deep venous thrombosis, pulmonary emboli and stroke [32-36]. Less frequently encountered side effects include benign ovarian cysts and ophthalmologic alterations [37]. The short-term toxic effects of tamoxifen may include vasomotor and gynecologic symptoms, such as vaginal discharge or irritation [38].

4. Classification of AIs and Mechanisms of Action

First-, Second- and Third-Generation Inhibitors

In postmenopausal women, AIs block the P450 cytochrome enzyme aromatase, responsible for the conversion of androgens to estrogens via a pathway that is the main source of estrogen, consequently suppressing estrogen levels. In postmenopausal populations, estrogens are produced in the adrenal glands, skin, muscles and adipose tissue [39]. In addition, a majority of breast tumors demonstrate the presence of intra-tumor aromatase activity, a likely source of local estrogen for tumor cells [40]. In contrast, for premenopausal women, AIs induce an increase in gonadotropin secretion secondary to the reduced negative feedback of estrogen to the pituitary. The consequence is ovarian stimulation and a potential increase in ovarian size, which may result in ovarian cysts, thereby confirming the absence of indication of AI treatment in premenopausal patients.

Effective AIs developed as therapeutic agents are described as first-, second- and third-generation inhibitors, according to the order of their clinical development. AIs were first developed as a non-surgical means to reduce estrogen production in patients with hormone-responsive tumors. The first-generation inhibitor is the nonsteroidal inhibitor aminoglutethimide (AG), which was the first AI to be studied in patients [41]. AG, an inhibitor of adrenal steroidogenesis, was studied almost 30 years ago. However, although significantly suppressing estrogen production, this agent was non-specific, altering as a consequence several intra-adrenal enzyme pathways, and producing sedative side-effects at the level of the central nervous system [42, 43], and reports of adrenal insufficiency led to its withdrawal from clinical use. AG blocks several cytochrome P450-mediated steroid hydroxylation steps, including those required for the cholesterol to pregnenolone conversion and for the aromatization of androgens to estrogens. However, AG is less specific and inhibits other cytochrome P450 enzymes involved in cortisol and aldosterone biosynthesis, which results in toxicity. Its efficacy in inhibiting aromatase activity stimulated the development of various new inhibitors during the 1980s and 1990s.

Subsequently, so-called 'second-generation' AIs, such as 4-hydroxyandrostenedione (formestane) and fadrozole (CGS 16,949A), were found to be significantly more potent and better tolerated than AG [44, 45]. Though more selective and potent compared to AG, fadrozole still has inhibitory effects on aldosterone, progesterone and corticosterone biosynthesis. On the other hand, formestane was the first selective AI to be used clinically, as it was effective and well tolerated [46]. However, formestane required intramuscular administration limiting its clinical use, and it did not show any benefit over tamoxifen [47].

In the late 1980s and early 1990s, research focused on developing agents with increased potency and higher selectivity. The identification of two mechanisms of aromatase inhibition led to the development of two types of third-generation AIs. Type I AIs are androgen analogues, which interfere with the substrate-binding site of the enzyme and block the enzymatic complex by producing an unbreakable covalent bond between the inhibitor and enzyme protein. Exemestane (Aromasin) is the only type I AI available for endocrine therapy. Nonsteroidal type II AIs block the electron transfer chain by the cytochrome P450 prosthetic group of the aromatase complex, acting as competitive inhibitors reversibly bound to the active enzymatic site. There are two type II AIs in clinical practice: Anastrozole (Arimidex) [48] and Letrozole (Femara) [49]. Today, these are widely used as first-line drugs in the endocrine treatment of hormone-dependent breast cancer.

5. Clinical Pharmacology of Third-Generation AIs

When compared with first- and second-generation inhibitors, third-generation inhibitors produce greater clinical benefits with near-complete specificity at clinical use. These drugs were also found to be better tolerated than tamoxifen and were associated with lower incidences of endometrial cancer, vaginal bleeding and discharge, cerebrovascular events, venous thromboembolic events, and hot flashes [50, 51]. In addition, the incidence of contralateral breast cancer occurrence was found to be significantly lower in the AI group than in the tamoxifen group [52, 53].

Secondary to good biodisponibility, all AIs (anastrozole, letrozole and exemestane) used clinically are given orally, once daily. The recommended daily doses are as follows: for anastrozole, 1 mg; for letrozole, 2.5 mg; and for exemestane, 25 mg. The time needed to reach maximal E_2 suppression is 2–4 days for anastrozole and letrozole and 7 days for exemestane [48, 54]. In a exemestane study, these measurements were performed every 7 days, and it is not known whether maximal estrogen suppression after exemestane may have occurred earlier. Anastrozole and exemestane achieve a steady-state drug level by 7 days [55]. In contrast, letrozole has been reported to take 60 days to achieve plasma steady-state levels, which may be reflective of the accumulation that occurs with letrozole over long-term dosing [56]. These data suggest a nonlinear relationship after repeated dose administration of letrozole. The half-lives of anastrozole, exemestane, and letrozole are 41 h, 27 h, and $\leq$4 days [57-59], respectively. All three AIs effectively reduce E_2, E_1, and E_1S: anastrozole by 81–94%, letrozole by 88–98% [60, 61]; and exemestane by 52–72% [55, 62]. In one prospective study, a comparative evaluation was carried out to determine the extent of plasma E_1, E_1S, and E_2 suppression after 6 weeks of treatment with either anastrozole or letrozole in postmenopausal women with metastatic disease [61]. There were no significant differences in reduction of plasma E_2 by either anastrozole (84.9%) or letrozole (87.8%). Anastrozole and letrozole reduced E_1 levels by 81.0% and 84.3%, respectively; the extent of suppression was significantly greater in the letrozole group ($P = 0.019$). Letrozole reduced plasma E_1S levels significantly more than anastrozole (98% vs. 93.5%, $P = 0.019$). The clinical significance of this additional suppression of estrogen by letrozole remains unclear. There are no comparative studies evaluating exemestane in relation to other agents, but one study reports that E_2 levels were suppressed by 65% with exemestane [55].

Breast cancer tumor cells and stromal cells have the potential to synthesize estrogen using local aromatase activity. Local production of estrogen by this mechanism results in a significantly higher concentration of estrogen in the tumor than in circulation in postmenopausal women. Separate studies evaluating these three agents have demonstrated that all three can reduce tumor estrogen [63-65]. As these studies were not carried out as comparative trials, it is difficult to assess whether one agent is more effective than another in reducing intratumoral estrogen. However, tumors have reduced 97–98% with all three AIs.

Initial second-line Phase III studies evaluated different doses of AIs and to determine whether higher doses would result in further reductions in intratumoral estrogen and in higher antitumor activity. Unfortunately, data from these studies did not provided convincing evidence for dose-dependent antitumor activities of AIs: the 1- and 10-mg doses of anastrozole had similar antitumor activity [66]; and second-line studies of 0.5 and 2.5 mg of letrozole [67] showed no consistent superiority of the higher dose. There are no comparable prospective studies in which different doses of exemestane have been studied.

6. Side Effects of AIs

Third-generation AIs (anastrozole, letrozole and exemestane) are replacing tamoxifen as adjuvant therapy in most postmenopausal women with early breast cancer. Although AIs have demonstrated superior efficacy and better overall safety than tamoxifen in randomized controlled trials, they may not provide the cardioprotective effects of tamoxifen, and bone loss may be a concern with long-term adjuvant use. Patients require regular bone mineral density monitoring, and prophylactic bisphosphonates are being evaluated to determine whether they may protect long-term bone health. Compared with tamoxifen, AIs decrease the risks of thromboembolic and cerebrovascular events, and the overall rate of cardiovascular events in patients treated with AIs is within the range seen in age-matched, non-breast-cancer populations. AIs are also associated with a lower incidence of endometrial cancer and fewer vaginal bleeding/discharge events than tamoxifen. , the incidence of hot flashes is also lower with anastrozole and letrozole compared with tamoxifen, but may be higher with exemestane. Generally, adverse events with AIs are predictable and manageable, whereas tamoxifen is associated with life-threatening events in a minority of patients.

6.1. Effects on Bone

The reductions in circulating estrogen levels that occur at menopause are associated with a rapid deterioration in bone mass by as much as 3% per year for the first 5 years after menopause [68]. After menopause, intrinsic production of estrogen from androgens largely occurs in nonovarian tissues, such as fat, muscle, skin and liver as a result of the activity of the aromatase enzyme complex. The resulting low levels of circulating estrogen help to prevent a more exaggerated bone loss [69]. AIs lower circulating estrogen levels to almost undetectable levels in postmenopausal women, predisposing them to bone loss with an increase in fracture risk. The rate of bone loss in women who experience premature menopause before the age of 45 or who undergo ovarian suppression therapy is accelerated by the concomitant use of AIs.

A randomized trial (Letrozole, Exemestane and Anastrozole Pharmacodynamics [LEAP]) of healthy volunteers demonstrated that letrozole, exemestane and anastrozole have similar effects on bone biochemical measurements and all result in increases in bone turnover [70]. There were no statistically significant differences among AIs from baseline to 24 weeks for B-ALP, serum C-telopeptide crosslinks, and propeptide of type I procollagen. The only difference in the bone remodeling markers was a greater decrease in parathyroid hormone with exemestane than with anastrozole ($P = 0.04$). In a recent study, bone health in 1,354 patients with breast cancer receiving AIs (anastrozole, exemestane and letrozole) was compared with 11,014 controls. Treatment with an AI increased the risk of bone loss (relative risk, 1.3; 95% CI, 1.1, 1.6; $P = 0.01$) and bone fracture (relative risk, 1.4; 95% CI, 1.2, 1.6; $P = 0.001$). The risks remained significantly higher for AI therapy after adjustment for age and co-morbidities [71]. An increase in the incidence of arthralgia is noted with all three AIs, compared with tamoxifen.

Tamoxifen was the first anti-estrogen drug shown to lower the risk of breast cancer recurrence. However, several large clinical trials have shown that AIs are more effective than tamoxifen at preventing the recurrence of estrogen-sensitive breast cancer. AIs also carry a lower risk of blood clots and endometrial cancer, both of which are possible side effects of tamoxifen. However, tamoxifen works differently in the body and can actually protect against bone loss. Postmenopausal women taking the drug anastrozole to prevent a recurrence of breast cancer suffered more bone loss than women taking tamoxifen, although anastrozole provided better protection against relapse, according to a substudy of a major clinical trial comparing the two drugs [72].

Women who take AIs to prevent a recurrence of breast cancer and who also have low bone density are sometimes also prescribed drugs called bisphosphonates. These drugs, which are FDA approved for prevention and treatment of osteoporosis (brittle bone disease), work by limiting further loss of bone density. Although the side effects of bisphosphonates are usually mild, in rare cases they can cause serious damage to the jaw or kidneys. In the American Society of Clinical Oncology (ASCO) guidelines, postmenopausal patients with breast cancer who receive AIs are identified as being at high risk for osteoporosis, and it is recommended that they have baseline bone mineral density (BMD) valuation and regular monitoring to guide subsequent therapeutic interventions such as bisphosphonates [73, 74]. In addition to bisphoshonates, denosumab is an investigational monoclonal antibody that inhibits a protein called RANK-ligand, which is involved in bone resorption. In a recent study, women treated with denosumab had a 5.5% increase in bone density in the lumbar spine compared with women treated with a placebo. This benefit occurred regardless of how long women had been taking AIs, the type of AI they were taking, or whether they had previously taken tamoxifen [75].

6.2. Effects on Lipids

Anastrozole, exemestane and letrozole have somewhat different effects on lipid profile. In one study of patients with metastatic disease ($n = 952$ at study entry), anastrozole showed no marked effects on lipid profile compared with baseline. Although these patients had advanced metastatic disease, these data provide some evidence that anastrozole did not significantly change the lipid profile compared with baseline data [76]. A smaller study ($n = 44$) that compared lipids at baseline and after 32 weeks of treatment with anastrozole [77] also showed no significant alterations in the lipid profile. However, in the ATAC trial, the incidence of hypercholesterolemia was higher in patients receiving anastrozole than in those receiving tamoxifen (9% vs. 3%, respectively; $P < 0.0001$) [78]. In the ITA trial, lipid metabolism disorders were reported in 9.3% of patients treated with anastrozole and 4.0% receiving tamoxifen ($P = 0.04$) [79].

Letrozole was evaluated in 20 women with advanced breast cancer who showed a significant increase in total cholesterol and low-density lipoprotein from baseline after 8 and 16 weeks of therapy [80]. In the BIG 1–98 trial, according to the protocol, cholesterol concentrations (fasting or non-fasting) were collected systematically in the case-report forms every 6 months and even patients with only a single measurement above the upper limit of normal were defined as hypercholesterolemic [81]. Hypercholesterolemia was reported in 5.4% of the letrozole arm, as compared with 1.2% of the tamoxifen arm, in patients with baseline values within normal limits, who then had an increase of 1.5 times the upper limit of normal [82].

Exemestane also decreased the total triglyceride levels in this study, and in a 9-week trial in advanced breast cancer, exemestane resulted in a significant decrease in both total cholesterol and HDL [83]. Although in another European Organization of Research and Treatment of Cancer study, 24 weeks of exemestane had no impact on the lipid profile [84], and in the IES trial of sequential exemestane after tamoxifen, hypercholesterolemia was not reported [52, 85].

6.3. Androgenic Effects

Anastrozole and letrozole have no androgenic, progestrogenic or estrogenic effects such as weight gain, acne or hypertrichosis. Exemestane has weak androgenic properties, and its use at higher doses has been associated with steroidal-like side effects such as weight gain and acne [62, 86]. These side effects have to be evaluated if exemestane is given on a prolonged basis, as in the adjuvant setting.

6.4. Effects on Adrenal Steroidogenesis

Data on anastrozole indicate that it has no impact on adrenal steroidogenesis at up to 10 times the clinically recommended dose, thus suggesting that this drug has little activity on other cytochrome P450 enzymes involved in steroid synthesis [87]. In one study in healthy postmenopausal women, basal and ACTH stimulation did not differ from baseline levels after 14 days of anastrozole at a 5- or 10-mg daily dose [87]. In another study, there were no changes in the ACTH-stimulated response after 115 days of therapy [88]. Letrozole studies showed a decrease in basal and ACTH-stimulated cortisol synthesis. In one study, patients with advanced breast cancer showed a significant decrease in ACTH-stimulated aldosterone levels after 3 months of letrozole treatment (2.5 mg or 0.5 mg; [89]). Exemestane in one small study had no impact on cortisol or aldosterone levels with up to 7 days of treatment, and the dose-ranging studies included doses from 0.5 to 800 mg [90]. Data from these limited studies demonstrate that anastrozole and exemestane have no effect on steroidogenesis. In contrast, letrozole alters cortisol and aldosterone levels even at therapeutic doses [55]. However, the clinical relevance of these differences remains to be defined.

6.5. Cardiovascular Effects

The rate of cardiovascular events in patients treated with AIs is well within the range seen in age-matched, non-breast-cancer populations; for example, for women 57–65 years of age, the rates of fatal myocardial infarction and other fatal coronary artery disease are 1.1 and 0.81 per 1,000 patient-years, respectively [91]. Similar rates were recorded in the UK General Practice Research Database and Swedish MI register. Currently, there is insufficient information to fully determine the effect of AIs on cardiovascular disease, particularly coronary heart disease.

6.6. Secondary Cancer

The association between tamoxifen and endometrial and uterine cancers is well established [24]. Meta-analysis has shown that tamoxifen is associated with a modest but statistically significant increase in the risk of developing gastrointestinal cancer (relative risk, 1.31; 95% CI, 1.01, 1.69), particularly for postmenopausal women (relative risk, 1.77) [92]. Although an association between cancer and AIs has not been observed, a recent safety analysis in the ATAC trial [78] showed a surprisingly higher incidence of head and neck cancer with anastrozole compared with tamoxifen (10/3092 vs. 3/3094, respectively). Similarly, there was an excess of lung cancer (25/3092 vs. 16/3094) and lung cancer deaths with anastrozole; however, further analyses are required to confirm these findings. Of note, a higher incidence of secondary cancer was not noted in the IES (72 events with exemestane vs. 107 with tamoxifen) or in the BIG 1–98 trial (69 events with letrozole vs. 82 with tamoxifen) [81, 85].

7. Other types of AI; Phytochemical Inhibitors

Studies have suggested that food sources, including cruciferous vegetables, soy, rye flour, grapes and mushrooms, are associated with a decreased risk of breast cancer. [93-97]. Bioactive components present in these food sources exert various biological activities and are being investigated for the prevention of both ER+ and ER– breast tumors. Some phytochemicals isolated from the plant kingdom, such as flavonoids and lignans, are known to be competitive inhibitors of aromatase, resulting in decreased levels of estrogen. For example, isolicoflavonol/prenylated flavonoid, isolated from the paper mulberry, nectandrin-B/lignan, isolated from *Myristicaceae*, Procyanidin B dimers, isolated from red wine and grape seeds, have been shown to be AIs, while *in vivo* studies in an aromatase-transfected MCF-7 breast cancer xenograft model also demonstrated that these chemicals reduced androgen-dependent tumor growth [96]. These results suggest that it is possible to generate potent new AIs that are derived from natural anti-aromatase chemicals. Efforts to develop new lines of AIs derived from these phytochemicals have been initiated in several laboratories.

8. Conclusions

Two different types of AI are clinically in use today; steroidal AIs, such as exemestane and non-steroidal AIs such as anastrozole and letrozole. The former binds permanently to aromatase and the latter binds reversibly to aromatase and compete with the precursors of estrogen. Both steroidal and non-steroidal AIs have been shown in large-scale clinical trials to be superior to tamoxifen in extending survival in women with metastatic disease, and in preventing recurrence when used as primary adjuvant therapy. In addition, treatment with an AI after a full course of tamoxifen continues to improve recurrence-free survival, as compared with cessation of hormone therapy.

The challenge remains to determine the best schedule for up-front AI treatment. Studies have shown that 5 years of AI alone is more effective at preventing recurrence than 5 years of tamoxifen alone, and that switching women already taking tamoxifen to an AI after 2 or 3 years has a more potent effect at preventing recurrence than continuing tamoxifen for a full 5 years. However, it is not yet clear whether sequencing tamoxifen and an AI in women who have not yet been treated with hormone therapy confers a benefit over 5 years of AI alone. Other questions remain as AIs move to a more prominent position in breast cancer treatment. AIs have their own side effects, most importantly loss of bone density, which can be particularly hazardous for women already at risk for osteoporosis. Therefore, tamoxifen may still provide a more favorable risk/benefit ratio for some subgroups of women, which need to be identified. In addition, the role of AIs in premenopausal patients remains to be defined. While AIs alone may not have an effect in premenopausal women, because the ovaries can override inhibition by producing larger amounts of aromatase, clinical trials are now testing AIs in premenopausal women in combination with drugs like goserelin, which suppresses ovarian function.

Because AIs have reduced the occurrence of contralateral breast cancer in several studies, researchers are now testing them as chemopreventive agents. Large-scale trials have begun testing exemestane and anastrozole in women at high risk for developing breast cancer. Although tamoxifen was approved in 1998 for the prevention of breast cancer in high-risk women, fewer women than expected have chosen to take it. If AIs prove to have both efficacy in preventing breast cancer occurrence and acceptable side effects, they and the new generation of a selective estrogen-receptor modulator (SERM) such as raloxifene will provide high-risk women with additional options to help prevent the disease.

Despite the wealth of data comparing AIs with tamoxifen, it is unclear whether anastrozole, exemestane and letrozole are clinically equivalent, owing to the lack of head-to-head trials directly comparing them. Preclinical and small clinical studies suggest that letrozole is the most potent inhibitor of aromatase, reducing circulating estrogen levels to a greater degree than the other agents. However, whether this greater activity translates into superior clinical efficacy remains to be determined. In the absence of direct comparative data, cross-trial comparisons have been used to gain insights into any safety or efficacy differences. All three AIs have been compared directly with tamoxifen, and their efficacy relative to tamoxifen has been compared across trials, yet such analyses are complicated by differences in treatment schedules, patient populations and trial designs. Although not definitive conclusions, some important differences are becoming clear, with upfront letrozole appearing particularly effective at preventing early distant metastasis, an event strongly associated with breast cancer-related death. No safety differences between these AIs have yet been identified. This article explores the pharmacologic and clinical differences between the AIs, based on data from clinical and preclinical studies.

In conclusion, the efficacy benefits of AIs outweigh the risks when they are used as adjuvant therapy in postmenopausal women with early breast cancer. Safety, QOL, and patient preference must all be considered in the determination of optimal strategies for long-term endocrine therapy, bearing in mind that patients may require treatment for 10 years or more. Every patient is unique, and endocrine therapy must be individualized according to clinical, biological, genomic and patient factors, such as lifestyle, the presence of significant co-morbidities and the use of concomitant medications.

References

[1] Deroo BJ, Korach KS. Estrogen receptors and human disease. *J Clin Invest.* 2006 Mar;116(3):561-70.

[2] Rochira V, Granata AR, Madeo B, Zirilli L, Rossi G, Carani C. Estrogens in males: what have we learned in the last 10 years? *Asian J Androl.* 2005;7(1):3-20.

[3] Shimodaira M, Nakayama T, Sato N, Saito K, Morita A, Sato I, Takahashi T, Soma M, Izumi Y. Association study of aromatase gene (CYP19A1) in essential hypertension. *Int J Med Sci. 2008 7;5:29-35.*

[4] MacDonald PC, Madden JD, Brenner PF, Wilson JD, Siiteri PK. Origin of estrogen in normal men and in women with testicular feminization. *J. Clin. Endocrinol. Metab.* 1979;49:905–916.

[5] McNatty, KP; Makris, A; DeGrazia, C; Osathanondh, R; Ryan, KJ. The production of progesterone, androgens, and estrogens by granulosa cells, thecal tissue, and stromal tissue from human ovaries in vitro. J Clin Endocrinol Metab. 1979;49:687–99.

[6] Ryan, KJ. Biological aromatization of steroids. J Biol Chem. 1959;234:268–72.

[7] Ryan, KJ; Naftolin, F; Reddy, V; Flores, F; Petro, Z. Estrogen formation in the brain. Am J Obstet Gynecol. 1972;114:454–60.

[8] Schindler, AE; Ebert, A; Friedrich, E. Conversion of androstenedione to estrone by human tissue. J Clin Endocrinol Metab. 1972;35:627–30.

[9] Harada, N; Sasano, H; Murakami, H; Ohkuma, T; Nagura, H; Takagi, Y. Localized expression of aromatase in human vascular tissues. Circ Res. 1999;84:1285–91.

[10] Mangelsdorf DJ, Thummel C, Beato M, Herrlich P, Schütz G, Umesono K, Blumberg B, Kastner P, Mark M, Chambon P, Evans RM. The nuclear receptor superfamily: the second decade. *Cell* 1995;83:835–839.

[11] Enmark E, Gustafsson JA. Oestrogen receptors - an overview. *J. Internal Med.* 1999;246:133–138.

[12] Levin ER. Cellular Functions of the Plasma Membrane Estrogen Receptor. Trends *Endocrinol. Metab*. 1999;100:374–377.

[13] Gruber CJ, Tschugguel W, Schneeberger C, Huber JC. Production and actions of estrogens. *N. Engl. J. Med.* 2002;346 (5):340–352.

[14] J.F. Couse and K.S. Korach, Estrogen receptor null mice: what have we learned and where will they lead us? *Endocrine Rev.* 1999;20:358–417.

[15] F.W. George and J.D. Wilson, Sex Determination and Differentiation. The Physiology of Reproduction, Knobil & Neill Raven Press 1988.

[16] M. Shozu, K. Akasofu, T. Harada and Y. Kubota, A new cause of female pseudohermaphroditism: placental aromatase deficiency. *J. Clin. Endocrinol. Metab.* 1991;**72** (3):560–566.

[17] N. Harada, H. Ogawa, M. Shozu and K. Yamada, Genetic studies to characterize the origin of the mutation in placental aromatase deficiency. *Am. J. Hum. Genet.* 1992;**51**:666–672.

[18] E.P. Smith, J. Boyd, G.R. Frank, H. Takahashi, R.M. Cohen, B. Specker, T.C. Williams, D.B. Lubahn and K.S. Korach, Estrogen resistance caused by a mutation in the estrogen-receptor gene in a man. *N. Engl. J. Med.* 1994;331:1056–1061.

[19] M. Faustini-Fustini, V. Rochira and C. Carani, Oestrogen deficiency in men: where are we today? *Europ. J. Endocrinol.* 1999;140:111–129.

[20] Zirilli L, Rochira V, Diazzi C, Caffagni G, Carani C. Human models of aromatase deficiency. *J Steroid Biochem Mol Biol.* 2008 ;109(3-5):212-218.

[21] Muti, P. The role of endogenous hormones in the etiology and prevention of breast cancer: the epidemiological evidence. *Recent Results Cancer Res.* 2005;166:245–256.

[22] Nolvadex Adjuvant Trial Organization Controlled trial of tamoxifen as adjuvant agent in management of early breast cancer. Interim analysis at four years by Nolvadex Adjuvant Trial Organisation. *Lancet.* 1983;1:257–261.

[23] Fisher B, Dignam J, Bryant J, DeCillis A, Wickerham DL, Wolmark N, Costantino J, Redmond C, Fisher ER, Bowman DM, Deschenes L, Dimitrov NV, Margolese RG, Robidoux A, Shibata H, Terz J, Paterson AH, Feldman MI, Farrar W, Evans J, Lickley HL. Five versus more than five years of tamoxifen therapy for breast cancer patients with negative lymph nodes and estrogen receptor-positive tumors. *J Natl Cancer Inst.* 1996;88:1529–1542.

[24] Fisher B, Costantino JP, Redmond CK, Fisher ER, Wickerham DL, Cronin WM. Endometrial cancer in tamoxifen-treated breast cancer patients: findings from the National Surgical Adjuvant Breast and Bowel Project (NSABP) B-14. *J Natl Cancer Inst.* 1994;86:527–537.

[25] Demissie S, Silliman RA, Lash TL. Adjuvant tamoxifen: predictors of use, side effects, and discontinuation in older women. *J Clin Oncol.* 2001;19:322–328.

[26] Meijer D, van Agthoven T, Bosma PT, Nooter K, Dorssers LC. Functional screen for genes responsible for tamoxifen resistance in human breast cancer cells. *Mol Cancer Res.* 2006;4:379–386.

[27] Fisher B, Dignam J, Bryant J, Wolmark N. Five versus more than five years of tamoxifen for lymph node-negative breast cancer: updated findings from the National Surgical Adjuvant Breast and Bowel Project B-14 randomized trial. *J Natl Cancer Inst.* 2001;93:684–690.

[28] Early Breast Cancer Trialists' Collaborative Group (EBCTCG). Effects of chemotherapy and hormonal therapy for early breast cancer on recurrence and 15-year survival: an overview of the randomised trials. *Lancet.* 2005;365:1687–1717

[29] Jones, KL; Buzdar, AU. A review of adjuvant hormonal therapy in breast cancer. *Endocrine-Related Cancer.* 2004; 11:392–406.

[30] Summaries for patients. Using medication to prevent breast cancer: recommendations from the United States Preventive Services Task Force. *Ann Intern Med.* 2002; 137:I62.

[31] van Leeuwen, FE; Benraadt, J; Coebergh, JW; Kiemeney, LA; Gimbrere, CH; Otter, R; Schouten, LJ; Damhuis, RA; Bontenbal, M; Diepenhorst, FW, et al. Risk of endometrial cancer after tamoxifen treatment of breast cancer. *Lancet.* 1994; 343:448–452.

[32] Fisher, B; Dignam, J; Wolmark, N; Wickerham, DL; Fisher, ER; Mamounas, E; Smith, R; Begovic, M; Dimitrov, NV; Margolese, RG; Kardinal, CG; Kavanah, MT; Fehrenbacher, L; Oishi, RH. Tamoxifen in treatment of intraductal breast cancer: National Surgical Adjuvant Breast and Bowel Project B-24 randomised controlled trial. *Lancet.* 1999; 353:1993–2000.

[33] Saphner, T; Tormey, DC; Gray, R. Venous and arterial thrombosis in patients who received adjuvant therapy for breast cancer. *J Clin Oncol.* 1991;9:286–294.

[34] Love, RR; Surawicz, TS; Williams, EC. Antithrombin III level, fibrinogen level, and platelet count changes with adjuvant tamoxifen therapy. *Arch Intern Med.* 1992; 152:317–320.

[35] Fisher, B; Costantino, JP; Wickerham, DL; Redmond, CK; Kavanah, M; Cronin, WM; Vogel, V; Robidoux, A; Dimitrov, N; Atkins, J; Daly, M; Wieand, S; Tan-Chiu, E; Ford, L; Wolmark, N. Tamoxifen for prevention of breast cancer: report of the National Surgical Adjuvant Breast and Bowel Project P-1 Study. *J Natl Cancer Inst.* 1998; 90:1371–1388.

[36] Dignam, JJ; Fisher, B. Occurrence of stroke with tamoxifen in NSABP B-24. *Lancet.* 2000; 355:848–849.

[37] Nayfield, SG; Gorin, MB. Tamoxifen-associated eye disease. A review. *J Clin Oncol.* 1996; 14:1018–1026.

[38] Love, RR; Cameron, L; Connell, BL; Leventhal, H. Symptoms associated with tamoxifen treatment in postmenopausal women. *Arch Intern Med.* 1991; 151:1842–1847.

[39] Miller WR, Dixon JM Endocrine and clinical endpoints of exemestane as neoadjuvant therapy. *Cancer Control.* 2002; 9:9–15.

[40] Bolufer P, Ricart E, Lluch A, Vazquez C, Rodriguez A, Ruiz A, Llopis F, Garcia-Conde J, Romero R. Aromatase activity and estradiol in human breast cancer: its relationship to estradiol and epidermal growth factor receptors and to tumor-node-metastasis staging. *J Clin Oncol.* 1992;10:438-446.

[41] Santen, R.J. Suppression of estrogens with aminoglutethimide and hydrocortisone (medical adrenalectomy) as treatment of advanced breast carcinoma: a review. *Breast Cancer Res. Treat.* 1981; 1: 183–202.

[42] Samojlik E, Veldhuis JD, Wells SA, Santen RJ. Preservation of androgen secretion during estrogen suppression with aminoglutethimide in the treatment of metastatic breast carcinoma. *J Clin Invest.* 1980; 65:602–612.

[43] Saphner T, Tormey DC, Gray R. Annual hazard rates of recurrence for breast cancer after primary therapy. *J Clin Oncol.* 1996; 14:2738-2746.

[44] Beretta KR, Hoeffken K, Kvinnsland S, Trunet P, Chaudri HA, Bhatnagar AS, Goldhirsch A, Cavalli F. CGS 16949A, a new aromatase inhibitor in the treatment of breast cancer--a phase I study. *Ann Oncol.* 1990; 1: 421–426.

[45] Dowsett M, Cunningham DC, Stein RC, Evans S, Dehennin L, Hedley A, Coombes RC. Dose-related endocrine effects and pharmacokinetics of oral and intramuscular 4-hydroxyandrostenedione in postmenopausal breast cancer patients. *Cancer Res.* 1989; 49: 1306–1312.

[46] Coombes RC, Goss P, Dowsett M, Gazet JC, Brodie A. 4-Hydroxyandrostenedione in treatment of postmenopausal patients with advanced breast cancer. *Lancet.* 1984;2: 1237–1239.

[47] Wiseman LR, McTavish D. Formestane. A review of its pharmacodynamic and pharmacokinetic properties and therapeutic potential in the management of breast cancer and prostatic cancer. *Drugs.* 1993;45:66–84.

[48] Plourde P. V., Dyroff M., Dowsett M., Demers L., Yates R., Webster A. ARIMIDEX: a new oral, once-a-day aromatase inhibitor. J. Steroid Biochem. *Mol. Biol.* 1995;*53*:175-179.

[49] Lipton A, Demers LM, Harvey HA, Kambic KB, Grossberg H, Brady C, Adlercruetz H, Trunet PF, Santen RJ.Letrozole (CGS 20267). A phase I study of a new potent oral aromatase inhibitor of breast cancer. *Cancer.* 1995; 75(8):2132-8.

[50] Baum M, Budzar AU, Cuzick J, Forbes J, Houghton JH, Klijn JG, Sahmoud T; ATAC Trialists' Group.Anastrozole alone or in combination with tamoxifen versus tamoxifen alone for adjuvant treatment of postmenopausal women with early breast cancer: first results of the ATAC randomised trial. *Lancet.* 2002; 359(9324):2131-9.

[51] Smith IE, Dowsett M. Aromatase inhibitors in breast cancer. *N Engl J Med.* 2003; 348(24):2431-42.

[52] Coombes RC, Hall E, Gibson LJ, Paridaens R, Jassem J, Delozier T, Jones SE, Alvarez I, Bertelli G, Ortmann O, Coates AS, Bajetta E, Dodwell D, Coleman RE, Fallowfield LJ, Mickiewicz E, Andersen J, Lonning PE, Cocconi G, Stewart A, Stuart N, Snowdon CF, Carpentieri M, Massimini G, Bliss JM; Intergroup Exemestane Study (2004) A randomized trial of exemestane after two to three years of tamoxifen therapy in postmenopausal women with primary breast cancer. N Engl J Med 350:1081–1092. Erratum in: *N Engl J Med.* 2004; 351:2461.

[53] Goss PE, Ingle JN, Martino S, Robert NJ, Muss HB, Piccart MJ, Castiglione M, Tu D, Shepherd LE, Pritchard KI, Livingston RB, Davidson NE, Norton L, Perez EA, Abrams JS, Therasse P, Palmer MJ, Pater JL. A randomized trial of letrozole in

postmenopausal women after five years of tamoxifen therapy for early-stage breast cancer. *N Engl J Med.* 2003; 349(19):1793-802.

[54] Demers L. M., Lipton A., Harvey H. A., Kambic K. B., Grossberg H., Brady C., Santen R. J. The efficacy of CGS 20267 in suppressing estrogen biosynthesis in patients with advanced stage breast cancer. J. Steroid Biochem. *Mol. Biol.* 1993; 44: 687-691.

[55] Zilembo N., Noberasco C., Bajetta E., Martinetti A., Mariani L., Orefici S. Endocrinological and clinical evaluation of exemestane, a new steroidal aromatase inhibitor. Br. J. *Cancer.* 1995; 72: 1007-1012.

[56] Bajetta E., Zilembo N., Dowsett M., Guillevin L., Di Leo A., Celio L., Martinetti A., Marchiano A., Pozzi P., Stani S., Bichisao E. Double-blind, randomised, multicentre endocrine trial comparing two letrozole doses, in postmenopausal breast cancer patients. *Eur. J. Cancer*. 1999; 35:208-213.

[57] Yates R. A., Dowsett M., Fisher G. V., Selen A., Wyld P. J. Arimidex (ZD1033): a selective, potent inhibitor of aromatase in postmenopausal female volunteers. *Br. J. Cancer*. 1996; *73:* 543-548.

[58] Spinelli R., Jannuzzo M. G., Poggesi I., Frevola L., Broutin F., Cicioni P. Pharmacokinetics (PK) of Aromasin (Exemestane, EXE) after single and repeated doses in healthy postmenopausal volunteers (HPV). *Eur. J. Cancer.* 1999; *35:*S295.

[59] Sioufi A., Sandrenan N., Godbillon J., Trunet P., Czendlik C., Howard H. Comparative bioavailability of letrozole under fed and fasting conditions in 12 healthy subjects after 2.5 mg single oral administration. *Biopharm. Drug Dispos.* 1997; 18:489-497.

[60] Geisler J., King N., Dowsett M., Ottestad L., Lundgren S., Walton P., Kormeset P. O., Lonning P. E. Influence of anastrozole (Arimidex), a selective, non-steroidal aromatase inhibitor, on in vivo aromatisation and plasma oestrogen levels in postmenopausal women with breast cancer. *Br. J. Cancer.* 1996; 74: 1286-1291.

[61] Geisler J., Anker G., Dowsett M., Lonning P. E. Letrozole suppresses plasma estrogen levels in postmenopausal breast cancer patients more completely than anastrozole. *Proc. Am. Soc. Clin. Oncol.* 2000; 19: 102a.

[62] Geisler J., Anker G., Dowsett M., Lonning P. E. Letrozole suppresses plasma estrogen levels in postmenopausal breast cancer patients more completely than anastrozole. *Proc. Am. Soc. Clin. Oncol.* 2000; 19:102a.

[63] Geisler J., Bernsten L., Ottestad B., Lindtjorn B., Dowsett M., Lonning P. E. Neoadjuvant treatment with anastrozole (Arimidex) causes profound suppression of intra-tumor estrogen levels. *Proc. Am. Soc. Clin. Oncol*. 1999; *18:*82a.

[64] Miller W. R. Biology of aromatase inhibitors: pharmacology/endocrinology within the breast. Endocr. Relat. *Cancer*. 1999; *6:*187-195.

[65] Dixon J. M., Grattage L., Renshaw L., Miller W. R. Exemestane as neoadjuvant treatment for locally advanced breast cancer: endocrinologic and clinical endpoints. *Breast Cancer Res. Treat.* 2000;*64:* 53.

[66] Jensen E. V., Block G. E., Smith S., Kyser K., DeSombre E. R. Estrogen receptors and breast cancer response to adrenalectomy. *Natl. Cancer Inst. Monogr.* 1971; 34: 55-70.

[67] Buzdar A., Douma J., Davidson N., Elledge R., Morgan M., Smith R., Porter L., Nabholtz J., Xiang X., Brady C. Phase III, multicenter, double-blind, randomized

study of letrozole, an aromatase inhibitor, for advanced breast cancer versus megestrol acetate. *J. Clin. Oncol.* 2001; 19: 3357-3366.

[68] Riggs BL, Khosla S, Melton III LJ. A unitary model for involutional osteoporosis: estrogen deficiency causes both type I and type II osteoporosis in post-menopausal women and contributes to bone loss in aging men. *J Bone Miner Res*. 1998; 13: 763–773.

[69] Cummings SR, Browner WS, Bauer D, Stone K, Ensrud K, Jamal S, Ettinger B. Endogenous hormones and the risk of hip and vertebral fractures among older women. *N Engl J Med.* 1998a; 339: 733–738.

[70] McCloskey E, Hannon R, Lakner G, Clack G, Miyamoto A, Eastell R. The letrozole (L), exemestane (E), and anastrozole (A) pharmacodynamics (LEAP) trial: a direct comparison of bone biochemical measurements between aromatase inhibitors (AIs) in healthy postmenopausal women. *J Clin Oncol*. 2006;24:16s.

[71] Mincey BA, Duh MS, Thomas SK, Moyneur E, Marynchencko M, Boyce SP, Mallett D, Perez EA. Risk of cancer treatment-associated bone loss and fractures among women with breast cancer receiving aromatase inhibitors. *Clin Breast Cancer*. 2006; 7:127–132.

[72] Eastell R, Adams JE, Coleman RE, Howell A, Hannon RA, Cuzick J, Mackey JR, Beckmann MW, Clack G. Effect of anastrozole on bone mineral density: 5-year results from the anastrozole, tamoxifen, alone or in combination trial 18233230. *J Clin Oncol.* 2008 Mar 1;26(7):1051-1057.

[73] Winer EP, Hudis C, Burstein HJ, Wolff AC, Pritchard KI, Ingle JN, Chlebowski RT, Gelber R, Edge SB, Gralow J, Cobleigh MA, Mamounas EP, Goldstein LJ, Whelan TJ, Powles TJ, Bryant J, Perkins C, Perotti J, Braun S, Langer AS, Browman GP, Somerfield MR. American Society of Clinical Oncology technology assessment on the use of aromatase inhibitors as adjuvant therapy for postmenopausal women with hormone receptor-positive breast cancer: status report 2004. *J Clin Oncol.* 2005; 23:619–629.

[74] Hillner BE, Ingle JN, Chlebowski RT, Gralow J, Yee GC, Janjan NA, Cauley JA, Blumenstein BA, Albain KS, Lipton A, Brown S. American Society of Clinical Oncology 2003 update on the role of bisphosphonates and bone health issues in women with breast cancer. *J Clin Oncol. 2003*; 21:4042–4057.

[75] Ellis GK, Bone HG, Chlebowski R, Paul D, Spadafora S, Smith J, Fan M, Jun S. Randomized trial of denosumab in patients receiving adjuvant aromatase inhibitors for nonmetastatic breast cancer. *J Clin Oncol.* 2008; 26(30):4875-82.

[76] Dewar J., Nabholtz J-M., Bonneterre J., Buzdar A., Robertson J. F., Thurlimann B. The effect of anastrozole (Arimidex) on serum lipids: data from a randomized comparison of anastrozole (AN) versus tamoxifen (TAM) in postmenopausal (PM) women with advanced breast cancer (ABC). *Breast Cancer Res. Treat.* 2000; 64: 51.

[77] Wojtacki J., Kruszewski W. J., Les'niewski-Kmak K., S'liwin'ska M., Czyzewska K., Kruszewska E. Short-term effects of anastrozole therapy on serum lipid profile in patients with breast cancer, previously treated with tamoxifen. Preliminary report. *Nowotwory,* 2001; 51: 43-47.

[78] The Arimidex, Tamoxifen, Alone or in Combination Trialists' Group, Buzdar A, Howell A, Cuzick J, Wale C, Distler W, Hoctin-Boes G, Houghton J, Locker GY, Nabholtz JM. Comprehensive side-effect profile of anastrozole and tamoxifen as adjuvant treatment for early-stage breast cancer: long-term safety analysis of the ATAC trial. *Lancet Oncol.* 2006; 7:633–643.

[79] Boccardo F, Rubagotti A, Puntoni M, Guglielmini P, Amoroso D, Fini A, Paladini G, Mesiti M, Romeo D, Rinaldini M, Scali S, Porpiglia M, Benedetto C, Restuccia N, Buzzi F, Franchi R, Massidda B, Distante V, Amadori D, Sismondi P. Switching to anastrozole versus continued tamoxifen treatment of early breast cancer: preliminary results of the Italian Tamoxifen Anastrozole Trial. *J Clin Oncol.* 2005; 23:5138–5147

[80] Elisaf M. S., Bairaktari E., Nicolaides C., Kakaidi B., Tzallas C. S., Katsaraki A. Effect of letrozole on the lipid in postmenopausal women with breast cancer. *Eur. J. Cancer.* 2001; 37: 1510-1513.

[81] Thürlimann B, Keshaviah A, Coates AS, Mouridsen H, Mauriac L, Forbes JF, Paridaens R, Castiglione-Gertsch M, Gelber RD, Rabaglio M, Smith I, Wardley A, Price KN, Goldhirsch A; Breast International Group (BIG) 1–98 Collaborative Group (2005) A comparison of letrozole and tamoxifen in postmenopausal women with early breast cancer. *N Engl J Med.* 2006; 353:2747–2757.

[82] Femara Prescribing Information (2005) Novartis.

[83] Engan T., Krane J., Johannessen D. C., Kvinnsland S. Plasma changes in breast cancer patients during endocrine therapy: lipid measurements and nuclear magnetic resonance (NMR) spectroscopy. *Breast Cancer Res. Treat.* 1995; 36: 287-297.

[84] Lohrisch C., Paridaens R., Dirix L. Y., Beex M., Nooij M., Cameron D. No adverse impact on serum lipids of the irreversible aromatase inactivator Aromasin [Exemestane (E)] in first-line treatment of metastatic breast cancer (MBC): companion study to a European Organization of Research and Treatment of Cancer (Breast Group) Trial with Pharmacia Upjohn. *Proc. Am. Soc. Clin. Oncol.* 2001; 20: 43a.

[85] Coombes RC, Kilburn LS, Snowdon CF, Paridaens R, Coleman RE, Jones SE, Jassem J, Van de Velde CJ, Delozier T, Alvarez I, Del Mastro L, Ortmann O, Diedrich K, Coates AS, Bajetta E, Holmberg SB, Dodwell D, Mickiewicz E, Andersen J, Lonning PE, Cocconi G, Forbes J, Castiglione M, Stuart N, Stewart A, Fallowfield LJ, Bertelli G, Hall E, Bogle RG, Carpentieri M, Colajori E, Subar M, Ireland E, Bliss JM; Intergroup Exemestane Study (2007) Survival and safety of exemestane versus tamoxifen after 2–3 years' tamoxifen treatment (Intergroup Exemestane Study): a randomised controlled trial. *Lancet* 2007; 17(369):559–570.

[86] Michaud L. B., Buzdar A. U. Risks and benefits of aromatase inhibitors in postmenopausal breast cancer. *Drug Saf.* 1999; 21:297-309.

[87] Plourde P. V., Dyroff M., Dukes M. Arimidex: a potent and selective fourth-generation aromatase inhibitor. *Breast Cancer Res. Treat.* 1994; 30:103-111,

[88] Buzdar A., Esparza-Guerra L. Anastrozole "Arimidex" does not impair adrenal cortisol or aldosterone synthesis in postmenopausal women with advanced breast cancer. Proc. Am. Soc. *Clin. Oncol.* 2001; 20:52b.

[89] Bisagni G., Cocconi G., Scaglione F., Fraschini F., Pfister C. H., Trunet P. F. Letrozole, a new oral non-steroidal aromatase inhibitor in treating postmenopausal patients with advanced breast cancer. *Ann. Oncol.* 1996; *7:* 99-102.

[90] Evans T. R., di Salle E., Ornati G., Lassus M., Benedetti M. S., Pianezzola E., Coombes R. C. Phase I and endocrine study of exemestane (FCE 24304), a new aromatase inhibitor, in postmenopausal women. *Cancer Res.* 1992; *52:*5933-5939.

[91] Sourander L, Rajala T, Raiha I, Makinen J, Erkkola R, Helenius H (1998) Cardiovascular and cancer morbidity and mortality and sudden cardiac death in postmenopausal women on oestrogen replacement therapy (ERT). Lancet 352:1965–1969. Erratum in: *Lancet.* 1999; 353:330.

[92] Braithwaite RS, Chlebowski RT, Lau J, George S, Hess R, Col NF. Meta-analysis of vascular and neoplastic events associated with tamoxifen. *J Gen Intern Med.* 2003; 18:937–947.

[93] Adlercreutz H. Phytoestrogens: epidemiology and a possible role in cancer protection. Environ *Health Perspect.* 1995; 103 Suppl 7:103-12.

[94] Glade MJ.Food, nutrition, and the prevention of cancer: a global perspective. American Institute for Cancer Research/World Cancer Research Fund, American Institute for Cancer Research, 1997. *Nutrition.* 1999; 15 (6):523-6.

[95] Grube BJ, Eng ET, Kao YC, Kwon A, Chen S.White button mushroom phytochemicals inhibit aromatase activity and breast cancer cell proliferation. *J Nutr.* 2001; 131(12):3288-93

[96] Eng ET, Ye J, Williams D, Phung S, Moore RE, Young MK, Gruntmanis U, Braunstein G, Chen S. Suppression of estrogen biosynthesis by procyanidin dimers in red wine and grape seeds. *Cancer Res.* 2003; 63(23):8516-22

[97] Kijima I, Phung S, Hur G, Kwok SL, Chen S. Grape seed extract is an aromatase inhibitor and a suppressor of aromatase expression. *Cancer Res.* 2006; 66(11):5960-7.

In: Aromatase Inhibitors: Types, Mode of Action and Indications ISBN: 978-1-60741-711-8
Editor: Jean R. Lamonte

Chapter II

Aromatase Inhibitors: Types, Mode of Action and Indications

Helena Aguilar and Ander Urruticoechea*
Institut Català d'Oncologia. Barcelona, Spain

Abstract

Tamoxifen, a selective estrogen receptor modulator (SERM), has been used for many decades as the "gold standard" endocrine treatment for patients with hormone-receptor-positive breast cancer. In the last few years, third generation aromatase inhibitors (AIs), either steroidal (exemestane) or non-steroidal (anastrozole, letrozole), have shown to be an effective alternative to tamoxifen in postmenopausal patients. These agents act blocking the aromatase enzyme which converts androgens into estrogens. They are associated with superior activity and better general tolerability when compared with tamoxifen both in the adjuvant and metastatic settings.

The potencial differences between steroidal aromatase inactivators and the non-steroidal aromatase inhibitors, manifested by the lack of total cross-resistance between compounds, underline the importance of exploiting the consequences of structural differences between aromatase inhibitors and inactivators.

A major obstacle to the clinical use of endocrine therapy is the occurrence of primary or acquired endocrine resistance. A better understanding of the multidirectional cross-talk between growth factor pathways, the estrogen receptors (both at the genomic and non genomic level), and estrogen receptor cofactors has shown to be fundamental to overcome or prevent resistance to endocrine therapy.

Thus, future research exploring the mechanisms of action and resistance to aromatase inhibitors will lead to the development of more effective therapeutic strategies.

* Author for correspondence: Translational Research Laboratory, Institut Català d'Oncologia, Av Gran via de L'Hospitalet 199-203, 08907 L'Hospitalet. Barcelona. Spain. Tel: 0034932607952; Fax: 0034932607466; anderu@iconcologia.net

Introduction

Endocrine treatment of hormone dependent breast cancer have tipically followed one of two estrategies. The deprivation of the estrogenic stimulus of breast cancer has been achieved via the withdrawal of estrogens through the ovarian ablation in premenopausal women or via the modulation of the estrogen receptor activity through pharmacological manipulation with tamoxifen. Other estrategies like androgens or anti-progestins have had limited clinical application. The introduction of tamoxifen as the first clinical effective antiestrogen more than 30 years ago, opened the era of contemporary endocrine therapy of breast cancer (Jordan, 2004). Tamoxifen has been the gold standard first-line therapy for the treatment of early and advanced breast cancer for about 30 years.

The concept of aromatase inhibition for the treatment of breast cancer was based on the historical finding by Santen et al., showing aminoglutethimide to inhibit the peripheral conversion of androgens into estrogens (Santen, Santner et al. 1978). Aminoglutethimide was soon established as second and third-line therapy in metastatic disease. The compound was not used as first-line therapy due to a comparable efficacy to tamoxifen and its side effect profile (Lonning and Kvinnsland 1988). During the recent 10–15 years, the role of tamoxifen was challenged by the development of more and more potent and selective aromatase inhibitors.

In contrast to the first-generation aromatase inhibitor aminoglutethimide (Jones, Powles et al. 1992; Boccardo, Rubagotti et al. 2001) the third-generation aromatase inhibitors; anastrozole, letrozole and exemestane, have revealed superiority with respect to relapse-free survival compared to tamoxifen in the adjuvant setting (Jakesz, Jonat et al. 2005; Coombes, Kilburn et al. 2007; Forbes, Cuzick et al. 2008). Third generation compounds showed superiority compared to tamoxifen as first-line treatment for metastatic cancer (Mouridsen, Gershanovich et al. 2001) and megestrol acetate as well as aminoglutethimide for second-line therapy (Dombernowsky, Smith et al. 1998; Kaufmann, Bajetta et al. 2000). In contrast, the results obtained with aminoglutethimide as well as the second-generation compounds fadrozole and formestane (4-hydroxyandrostenedione) revealed clinical efficacy for these compounds similar to tamoxifen and megestrol acetate. Thus, the third-generation aromatase inhibitors differ from their predecessors with respect to treatment efficacy (Lonning 2002) and to a much more favourable toxicity profile. The compounds may be divided into two distinct classes; anastrozole and letrozole are both non-steroidal compounds (imidazole derivatives), in contrast with the steroidal structure of exemestane, a derivative of the aromatase substrate androstenedione (Figure 1). The mechanism of action of AIs differs substantially from tamoxifen. Whereas tamoxifen produces a blockade in the ability of E2 to induce the activation of ER, AIs simply produce an almost complete deprivation of the substrate of the reaction, E2, mediated by the interference with the low, peripheral, activity of the aromatase enzyme typical of the postmenopausal stage. The increased efficacy of aromatase inhibitors over tamoxifen therapy has recently been demonstrated by clinical trials, a significant increase in disease free-survival has been shown using three generation aromatase inhibitors (Coombes, Hall et al. 2004; Goss, Ingle et al. 2005; Howell, Cuzick et al. 2005). All AI are similar in that they inhibit estrogen synthesis by blocking aromatase activity, however, there are distinct differences among them. These are either class-independent differences in terms of potency (Geisler and Lonning 2005) or class-specific differences in terms of the mechanism of binding to aromatase. There may also be clinically relevant differences among AIs related to the androgenic properties of esteroidal AIs.

The endocrine dependency of breast epithelial cells is key for the treatment of breast cancer. Oestrogen levels in breast cancer of menopausal women are about 20-fold higher than levels in the plasma (Pasqualini, Chetrite et al. 1996). Different strategies of oestrogen deprivation range from ovarian ablation in pre-menopausal women to the new aromatase inhibitors (anastrozol, letrozole and exemestane) including ER down regulators (fulvestrant), selective oestrogen receptor modulators (tamoxifen, raloxifen and toremifen), LHRH analogues, androgens, progestagens and high doses of oestrogens have been used. For years, selective estrogen modulators (SERMs) have been vitually the only effective pharmacological approach to endocrine treatment of breast cancer. Tamoxifen is a dual agonist-antagonist that binds to ER and produces a conformacional change in the receptor blocking the binding of co-activators (Shiau, Barstad et al. 1998). This dual antagonistic-agonistic effect is responsible for several side-effects of the drug, some of them desirables (delay in bone resorption in post-menopausal women) and others more damaging (endometrial hyperplasia and cancer). The benefits of adjuvant tamoxifen treatment are independent of age and are restricted to women with ER positive breast cancer (Osborne 1998) following tumor excission.

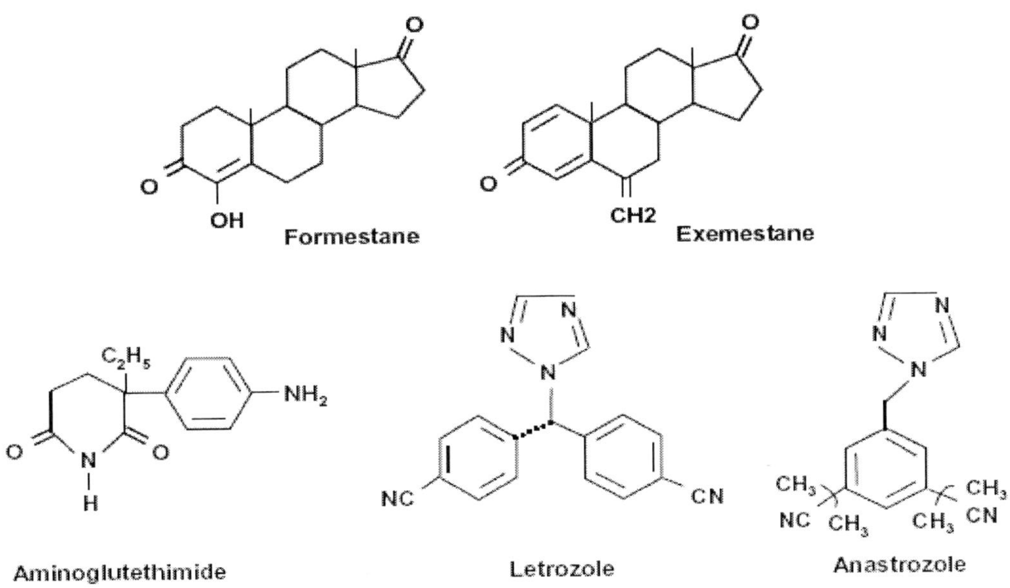

Figure 1. Structures of the non-steroidal first generation agent aminoglutethimidine, the second-generation steroidal compound formestane and three third-generation aromatase inhibitors.

The conversion of androgens into oestrogens in pre-menopausal women takes place mainly in the ovarian cells surrounding the follicles. During this period the aromatase produces high levels of circulating E2 and the AIs cannot significantly decrease the ovarian aromatase activity being of little use in pre-menopausal patients with breast cancer. Conversely, when the pool of follicles comes to an end and women undergo menopause, the ovary stops oestrogen production. Thereafter, E2 levels are much lower but are still present. Although E2 production continues thanks to aromatase activity, this no longer occurs in the ovaries but rather in several tissues incuding fat tissue, adrenal glands, muscles, brain, breast parenchyma and breast cancer itself when present. The activity of aromatase is substantially lower after menopause and may be blocked by AIs, achieving aromatase inhibition rates over 95% (Geisler, King et al. 1996; Geisler, King et al. 1998; Geisler, Haynes et al. 2002). Thus E2 are reduced to a minimum leaving ER without its natural substrate and stopping ER lead gene transcription.

The Oestrogen Pathways

It has been well established that the activity of all oestrogen-sensitive cells is mediated by the binding of E2 to ER□(McGuire, 1972). In the "classical" model of ligand-activated transcription, the estrogen passively diffuses throught the cell membrane an binds to ER located in the nucleus. After binding its ligand, ER dissociates from heat shock proteins, and undergoes conformational changes, phosphorylates and dimerizes with another ER. The activated ER-dimers release co-repressor, recruit co-activator proteins (Normanno, Di Maio et al. 2005) and these activated complexes bind to oestrogen response elements (EREs) along the DNA sequence producing the transcription of genes involved in proliferation, inhibition of apoptosis and the promotion of angiogenesis, invasion and metastasis among others. Nevertheless, this classical genomic mechanism of action has recently been enriched by other alternative modes of action of the complex E2-ER. These different pathways can be divided into classical genomic, non-classical genomic and non-genomic. One-third of genes regulated by ER do not contain ERE-promotor sequences. The AP-1 binding sites are DNA sequences where ER bind, and also result in gene activation/repression. This is known as the "non-classical" genomic mechanism of action. This AP-1 element binding depends on other transcription factors such as JUN and FOS. A relatively recently described mechanism by which the ER can influence cell proliferation and the rest of cancer cell processes is throught "non genomic" molecular cross-talk with tyrosine receptor kinase growth factor receptors and their downstream effector molecules (Kushner, Agard et al. 2000). The estrogen receptor has been associated with the cytoplasm and cell membrane trough its association with membrane anchors (Jacob, Sebastian et al. 2006). In this way, ER can directly interact with the pathways depending on insulin-like growth factor receptor (IGFR-1), Src, phospatidylinositol 3-kinase (PI3K), mitogen-activated protein kinase (MAPK), epidermal growth factor receptor (EGFR) and ErbB-2 (Filardo, Quinn et al. 2000; Kahlert, Nuedling et al. 2000). Conversely, the kinases can modulate ER activation via the phosphorylation of co-factors that activate or repress ER activity on transcription regulation (Bunone, Briand et al. 1996; Campbell, Bhat-Nakshatri et al. 2001). Finally, in the scenario of the non-genomic interactions of ER, it has been shown that ER can be directly phosphorylated by the kinases resulting in its ligand-independent activation. This cross-talk has great importance in the modulation of hormonal treatments and, particularly, in the acquisition of resistance to such therapy.

Transcriptional profile of estrogen responses in ER+ human breast cancer cell lines and model systems in vitro shows differents changes in the transcription of large number of genes

(Frasor, Danes et al. 2003; Frasor, Stossi et al. 2004), but very little is known of these effects in vivo and whether these molecular changes can be determinants of clinical response. With the advent of microarray technology, the effects of aromatase inhibitors on gene expression can be compared on a very large number of genes without any prior selection bias. In fact, gene expression profiling in breast cancer using microarrays has been carried out in a number of studies investigating genes overexpressed in breast cancer (Jiang, Harlocker et al. 2002) or genes associated with clinical outcome or prognosis (van 't Veer, Dai et al. 2002; van de Vijver, He et al. 2002; Huang, Cheng et al. 2003), response to chemotherapy (Sotiriou, Powles et al. 2002), tumor aggressiveness (Zajchowski, Bartholdi et al. 2001), or classification of primary tumors (Perou, Jeffrey et al. 1999; Perou, Sorlie et al. 2000; Sorlie, Perou et al. 2001). Several studies have also investigated gene expression patterns associated with the ER status of breast cancers (Dressman, Walz et al. 2001; West, Blanchette et al. 2001) and biological response to AIs itself (Mackay 2007)

Mechanism of Action of Aromatase Inhibitors

The AIs act through inhibition of the cytocrome P450 (CYP) enzyme aromatase which catalyzes the conversion of androgens into oestrogens (di Salle, Ornati et al. 1992; Johnston and Dowsett 2003; Brueggemeier, Hackett et al. 2005). The non-steroidal type II compounds, such as anastrozole and letrozole, are characterized by the imidazole/triazole structure and interact non covalently with the enzyme preventing binding of the androgens to the catalytic site (Cole and Robinson 1990). This antagonism is reversible and the type II AIs can be competitively displaced from the active site by endogenous substrate. In contrast, the steroidal type I inhibitor exemestane is an analogue of the natural aromatase substrate androstenedione. This compound is recognized by the active site of aromatase as an alternative substrate (di Salle, Ornati et al. 1992; Brueggemeier, Hackett et al. 2005; Hong, Yu et al. 2007) and binds irreversibly and covalently to the substrate-binding site of aromatase, permanently inactivating the enzyme.

Diferences in binding between nonsteroidal and steroidal AIs lead to divergent effects on aromatase in experimental systems. In MCF-7aro breast cancer cell line, that has been transfected with the aromatase gene, exemestane destabilizes the enzyme and then is degraded by proteosomic enzymes (Wang and Chen 2006). Non steroidal inhibitors may increase aromatase protein levels because they stabilize the structure of the enzyme and/or induce transcription of aromatase mRNA (Chen, Zhou et al. 1999; Kao, Okubo et al. 1999). This upregulation, together with the reversible nature of type II results in increased enzyme activity following exposure to these drugs. This is consistent with data from ex vivo assays for aromatase activity following neoadjuvant treatment with aminoglutethimide, a potent inducer of multiple CYP enzymes (Miller and O'Neill 1987; Lonning 1990).

There appears to be only one human aromatase gene and a single protein translate, although aromatase in normal tissues is differentially regulated through a variety of promoters (Bulun, Takayama et al. 2004). However, it is possible that in tumors there may be mutations that could affect aromatase protein structure or sensitivity to inhibitors. In this respect, structural functional studies on aromatase have produced proteins that appear resistant to a steroidal AI (formestane) while maintaining sensitivity to nonsteroidal AIs (Miller, Anderson et al. 2003). Such a phenotype has been reported in some breast cancers, for which causative mutations have not been identified (Sourdaine, Parker et al. 1994). The possibility exists, therefore, that aromatase in individual breast cancers may be more susceptible to one class of AI than another.

While aromatase protein responds to one only gene sequence ten gene promoters lead to a high complexity in its regulation. With exception of a single promoter (pII) located at the beginning of exon II, these promoters all are located within the untranslated exon I. They get stimulated by different ligands and are differentially used across the different tissue compartments. Aromatase expression is mainly regulated through the exon 1.4 promoter in the normal breast tissue and there is a regulatory shift during the malignant transformation (Bulun, Sebastian et al. 2003; Chen, Ye et al. 2005) because aromatase expression in breast cancer is mainly regulated by the 1.3 and PII. Only four polimorphisms are located in the coding regions and three of them are associated with a modest reduction in protein expresion (Ma, Adjei et al. 2005).

The androgenic structure of type I steroidal AIs may give rise to hormonal effects apart from the decrease in estrogen production caused by inhibition of aromatase. When considering the androgenic properties of AIs, differential effects on healthy tissue may also affect clinical activity for breast cancer patients. Work in model systems suggests that exemestane treatment may lead to fewer adverse effects related to bone loss than nonsteroidal AIs because of its androgenicity (Goss, Hadji et al. 2007; Miki, Suzuki et al. 2007). Preclinical studies have demonstrated that these effects are related to activation of the AR by exemestane in osteoblasts (Miki, Suzuki et al. 2007). Several preliminar clinical findings support a lower degree of bone loss with exemestane (Lonning, Geisler et al. 2005; Coleman, Banks et al. 2007; Goss, Qi et al. 2007) although this hypothesis remains untested in large randomized trials.

Lack of Cross-resistance

An issue of particular interest is the lack of complete cross-resistance between steroidal and non-steroidal compounds. Although the studies reporting this phenomenon in general contain a small number of patients, some general conclusions may be drawn. Several potential mechanisms have been suggested including differential sensitivities of aromatase variants to specific compounds, the nature of the interactions with the enzime's active site; androgen-agonistic effects; effective uptake in tumor tissue and inherent differences in potencies among AIs.

The first demostration of the lack of cross-resistance between different AIs was reported in advanced breast cancer patients who showed clinical response to formestane after progression on aminoglutethimidine (Murray and Pitt 1995). Some studies demonstrated that patients exposed to non esteroidal AIs may benefit from a steroidal compound of similar biochemical potency. Although the objective response rate is low, durable stable disease is achieved in about 20% of patients treated for up to 6 months. Observations based on small studies demonstrated a response to anastrozole after formestane and to letrozole after exemestane.

Recently, indirect evidence in favour of lack of complete cross-resistance between third generation non-steroidal aromatase inhibitors (anastrozole and letrozole) and exemestane was indirectly substantiated by the EFECT study (Chia, Gradishar et al. 2008) in which patients with advanced hormone-responsive breast cancer refractory to a nonsteroidal AI were randomized to receive either fulvestrant or exemestane. The objective response rate as well as stable disease was similar in the two arms. In addition, no difference was recorded with respect to time to progression either.

Clinical Guidelines for the Use of Aromatase Inhibitors

Following the traditional guidelines for the clinical development of new treatments, AIs (initially letrozol and anastrozol) were tested in the advanced (metastatic or locally advanced) setting of breast cancer. When compared to the standard, tamoxifen, AIs showed a modest benefit in terms of response rate and overall survival. As a matter of fact the only aromatase inhibitor to show increased percentage of tumor shrinkage compared to tamoxifen was letrozol. See table I.

Table 1. Tamoxifen versus AIs in first line metastatic or locally advanced breast cancer

Author	Treat.	n	Response rate	Clinical benefit rate	TTP (months)
Mouridsen	Letrozole	453	30%	49%	9.4
	Tamoxifen	454	20%	38%	6.0
Nabholtz	Anastrozole	171	21%	59%	11.1
	Tamoxifen	182	17%	46%	5.6
Bonneterre	Anastrozole	340	33%	56%	8.2
	Tamoxifen	328	33%	56%	8.3
Eiermann	Letrozole	154	55%	--	--
	Tamoxifen	170	36%	--	--
Ellis	Letrozole	154	60%	--	--
	Tamoxifen	170	41%	--	--

Underlined values show a statistically significant benefit in favour of AIs. TTP: time to progression.

On the basis of the significantly increased clinical benefit, letrozol is now the new standard of care for postmenopausal patients candidate for endocrine treatment of metastatic breast cancer.

As part of its clinical development letrozol showed a remarkably increased response rate when compared to tamoxifen in the primary, neoadjuvant setting. In this situation, hormonal treatment is given for some months prior to surgery in order to allow large or locally advanced tumors to be deemed amenable for surgery. These differences where particularly evident when tumors with up-regulated Her-1 and Her-2 genes were considered Ellis 2001.

Following type 2 AI failure in the advanced setting, tamoxifen remains second line standard. When tumor become resistant to letrozole or anastrozole and tamoxifen and on the basis of the trials substantiating the lack of total cross-resistance between steroidal and non-steroidal AIs (Chia, Gradishar et al. 2008) the third line endocrine treatment is either exemestane or fulvestrant.

Endocrine adjuvant treatment for hormone receptor positive early breast cancer following tumor excision has been a busy clinical trial battlefield for the last ten years. Tens of thousand of patients have been randomized to different strategies of administration of AIs alone or in combination with tamoxifen in an enormous effort of the manufacturers of letrozol, anastrozol and exemestane to prove their drug to be the best in this setting. Following all these investigations no single strategy has emerged as a single standard, nevertheless some general recommendations (summarized in national or international guidelines) (Winer, Hudis et al. 2005). Firstly, tamoxifen for five years is no longer the standard endocrine treatment for postmenopausal women with hormone receptor positive tumors following surgical excision. At some point of the adjuvant treatment strategy, aromatase inhibitors have to be introduced. The different therapeutic schemes include: AI for five years (no tamoxifen), AI for three to five years following two years of tamoxifen and AI for two to three years following five years of tamoxifen.

The decision on which particular regimen to yours depends greatly upon the differential toxicity profile of tamoxifen and AIs and upon patient and physician preferences.

Resistance to Aromatase Inhibition

Although AI treatment has shown to be effective in the clinic, resistance to these therapies still occurs and is highly problematic, due to the lack of response to current endocrine therapy. There are two types of endocrine resistance. De novo/intrinsec resistance refers to lack of response at initial exposure to endocrine therapy of aromatase-positive and estrogen receptor positive breast cancers. The most obvious mechanism of acquired resistance involves a selection process. All tumors are heterogeneus. All breast tumors has ER+/estrogen responsive and ER- or estrogen independent cells. However, clinically, many of the resistant tumors are still ER positive. The mechanisms of adquired resistance are probably similar to the novo/intrinsec resistance with the diference that the adquired resistence is developed during endocrine therapy of patients who respond to the treatment inicially. It is very unlikely that acquired resistance results from aromatase or ER mutation developed during the treatment. This resistance is the adaptative response of the cells resulting from a crosstalk between ER and growth factor pathways or other currently identified pathways (Dowsett 1996; Lonning 1999; Jordan 2003; Schiff, Massarweh et al. 2003).

Preclinical and clinical findings support the hypothesis that a major mechanism of resistance to endocrine therapy is he adquisition by breast cancer cells of an increased sensitivity to estrogen. Extensive work on estrogen deprivated MCF-7 cells has revealed that these cells adapt by developing estrogen hypersensitivity achieving maximal growth stimulation at E_2 concentrations of 10^{-4}–10^{-5} times the concentration needed for the mother cell line (Masamura, Santner et al. 1995; Martin, Farmer et al. 2003). All mechanisms involved in the process of adaptation to low estrogen levels are not fully understood, but several signalling pathways like the mitogen-activated protein kinase (MAPK) pathway have been suggested to play a major role. Santen's group has established an in vitro model of long-term estrogen deprivation LTED by culturing MCF7 cells in estrogen free medium (Santen, Song et al. 2005). The hipersensitivity of these cells was confirmed by the ability of estrogen to stimulate their proliferation at concentrations four times lower as compared with wild-type MCF-7 cells. In this model, the non-genomic ER□ funtions are significantly enhanced and growth factor and PI3K/AKT pathways are significantly increased. The activation of MAPK and AKT has been associated with resistance to endocrine therapy and a worse outcome in breast cancer patients. On the other hand, Dowsett's group highlighed the important role of the crosstalk with ErbB2 instead of IGF-IR signaling in the process of resistance acquisition in LTED cells (Chan, Martin et al. 2002). After prolonged culture in the absence of estrogen ERa functions independently of exogenous estradiol. These cells show increased levels of phosphorilation of ERa at serine118, a site that can be phosphorilated by different intracellular kinases. Other model of resistance to estrogen withdrawal, MCF-7X, has been developed by Nicholson and co-workers. These cells have an estrogen dependent ER□ pathway and their growth is mainly supported by the PI3K/AKT pathway. In these cells no activation of EGFR/ErbB-2 or IGF-1R was demonstrated (Osborne and Schiff 2003; Schiff, Massarweh et al. 2003; Schiff, Massarweh et al. 2004). The differences in the mechanism of resistance of these models may account for differences of culture conditions, for example MCF7-X are cultured under estrogen and growth factors deprivation conditions.

Overexpression of HER-2 in about 20% of all breast cancers has been linked to reduced responsiveness to tamoxifen, but not to aromatase inhibition. The small subset of ER positive tumors that also overexpressed the growth factor receptors HER2 and/or epidermal growth factor receptor (EGFR) are relatively resistant to tamoxifen and more sensitive to letrozole. The clinical finding that aromatase inhibitors are superior to tamoxifen in HER2-neu positive patients in the neoadjuvant settings suports the theory that tamoxifen may act as an agonist in breast cancer cells with overactive growth factor receptor signaling networks (Ellis, Coop et al. 2001; Smith, Dowsett et al. 2005). A recent retrospective analysis of the BIG 1-98 trial suggested that HER2 overamplification was associated with a high relapse rate, regardless of whether tamoxifen or letrozole was used (Mauriac, Keshaviah et al. 2007).

Recently, a role for nongenomic ER activity in mediating estrogen-induced growth and survival effects has also been described in breast cancer cells, and the existance of membrane ER has now been established by numerous biochemical, immunohistological and genetic methods. This membrane ER can potentially interact with and activate via phosphorilation several membrane tyrosine kinases in breast cancer cells. Estrogen receptor can directly interact with the insulin-like growth factor receptor (IGFR) producing an activation of IGFR downstream activation. It can also interact with HER2 in the membrane and estrogen-activated membrane ER can phosphorylate and activate the EGFR in a process that involves activation of G-proteins, c-Src and MMPs (Razandi, Pedram et al. 2003). All this increases signaling of both ER and growth factor receptor pathways and enhances propagation and survivability of a breast cancer cell by the contribution of multiple pathways making these cells more resistant to single forms of molecular therapy.

Strategies to Overcome Resistance to Aromatase Inhibition

On the basis of the previous evidence supporting the key role of growth factor signaling in resistance to endocrine treatments, the most frequently explored mean to overcome resistance is the combination of growth bfactor signal inhibitors plus hormonal therapies. As detailed before breast cancer cells that become resistant to endocrine therapy with either tamoxifen or estrogen deprivation have been shown to be characterized by enhanced expression of EGF-R/HER-2 and subsequent MAPK activation (Schiff, Massarweh et al. 2003). Another pathway that may be activated in hormone-resistant cells is the PI3K/Akt pathway (Johnston, Head et al. 2003; Johnston, Martin et al. 2005). Akt promotes cell survival and is activated by many different growth factors like insulin like growth factor (IGF), basic fibroblast growth factor (bFGF), epidermal growth factor (EGF), and vascular endothelial growth factor (VEGF).

The results of preclinical studies that use signal transduction inhibitors in combination with endocrine therapy shown additive or sinergistic effects. As an example these effects have been demonstrated when farnesyl transferase inhibitors are combined with tamoxifen or AIs in ER-positive breast cancer cells (Ellis, Vos et al. 2003; Johnston, Head et al. 2003). A synergistic anti-tumor efect have been reported for trastuzumab when combined with tamoxifen in ER-positive ErbB-2-over-expressing BT474 breast cancer cells, although this combination did not induce apoptosis (Argiris, Wang et al. 2004). Combined treatment of MCF-7 cells with tamoxifen and the EGFR-tyrosine kinase inhibitor gefitinib was more effective in inhibiting proliferation, promoting apoptosis and eliminating bcl-2 as compared with tamoxifen alone (Gee, Harper et al. 2003). Finally, mammalian target of rapamycin (mTOR) acts as a downstream effector of the PI3K/Akt pathway regulating transition through the G1/S phase of the cell cycle. Several approaches blocking selected pathways have been developed. This includes the PI3K-inhibitor LY294002 and rapamycin analogues like temsirolimus (CCI-779) or everolimus (RAD-001). Additive effects were obtained with combinations of mammalian target of rapamycin (mTOR) antagonists and estrogen deprivation therapy with letrozole in preclinical models (Boulay, Rudloff et al. 2005).

The individual targets of these novel treatment strategies are summarized in Fig. 2. While monotherapy with these novel compounds has revealed low response rates (Johnston, Martin et al. 2005), clinical trials have been initiated combining aromatase inhibitors with monoclonal antibodies, tyrosine kinase inhibitors, farnesyltranferase inhibitors or m-TOR inhibitors. Many of these trials are designed for the first-line treatment in the metastatic setting.

A number of phase II studies have evaluated the combination of transtuzumab and AI therapy in the first and second line setting. Preliminary promising results led to the initiation of a phase III trial examining the combination of anastrozole with transtuzumab versus anastrozole alone in women with metastatic ER and/orPR positive, with evidence of HER2 overamplification. This study showed the superiority of the combination arm although the low response rate showed in both arms blurred the clinical relevance of the positive results. Studies in the future are addressed to investigate the role of pertuzumab, an antibody that blocks HER2 dimerization and has been shown in vitro to promote the antitumor efficacy of fulvestrant. Other strategy that is under study is the use of tyrosine kinase inhibitors in combination with endocrine therapy. One of these studies included 53 untreated postmenopausal patients with ER and EGFR positive primary breast cancer randomizing them either to 4 weeks of gefitinib alone or gefitinib plus anatrozole in a controlled double blind trial and showed that the combination of gefitinib plus anastrozole was more effective reducing Ki-67 than gefitinib alone (Polychronis, Sinnett et al. 2005).

The major question is, whether combined therapy of an STI with traditional endocrine therapy might prolong time to disease progression (TTP) by delaying endocrine resistance. The results of these trials will be necessary to define particular subgroups of patients that may profit from these novel drugs.

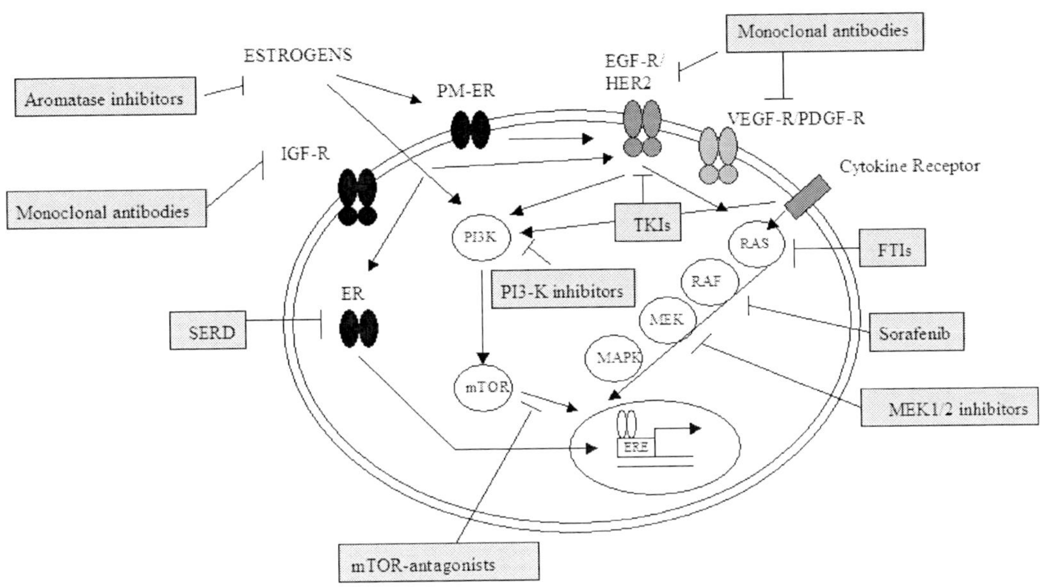

Figure 2. Cross-talk between growth factors, cytokines and hormones in endocrine breast cancer.

Alternative strategies are the combinations of steroidal antiestrogens with aromatase inhibitors. Fulvestrant (Faslodex®) is a novel, steroidal ER antagonist, preventing dimerisation of the ER (Howell, Osborne et al. 2000). Due to rapid degradation of the fulvestrant-ER-complex, cells experience a decrease in cellular ER. Fulvestrant is lacking estrogen agonistic effects and has been shown to be effective in patients resistant to tamoxifen therapy (Howell, Robertson et al. 2004). In clinical phase III trials, enrolling postmenopausal women with advanced disease, progressing on first-line therapy with tamoxifen, fulvestrant was comparably effective to anastrozole (Osborne, Pippen et al. 2002; Robertson, Osborne et al. 2003). The rationale to combine fulvestrant with aromatase inhibitors has been generated from preclinical studies (Johnston, Martin et al. 2005). It has been hypothesized that increasing levels of estradiol added to tamoxifen-resistant cells probably prohibit effective ER-antagonism by fulvestrant. Based on these preclinical studies with fulvestrant, several phase II and III trials involving the aromatase inhibitor anastrozole and the aromatase inactivator exemestane have been initiated. These ongoing trials will help to improve sequencing of endocrine therapy in the metastatic setting.

Conclusion

The introduction of aromatase inhibitors in the therapeutic scenario of hormone receptor positive breast cancer has led to a significant improvement in the outcome of patients with this disease. This has been the case both in the adjuvant setting after the excision of early, operable, breast cancer where the overall and disease free survival rates have significantly increased and in the metastatic and non-resecable disease where progression free survival times are now prolonged with a low impact of treatment related toxicity.

On top of this, the lack of complete cross-resistance between the two families of third generation AIs has provided clinicians with a valuable tool to prolong the time of disease control, in particular in the palliative setting.

Nevertheless, and irrespectively of the evident benefit provided by AIs, the general landscape of hormonal treatment in breast cancer remains the same. This way, and whenever tumor excision is not possible, resistance to endocrine maneuvers is virtually universal. The onset of this resistance is variable, ranging from up-front growth of the tumor irrespective of the initiation of treatment to the late onset of acquired resistance.

In all cases, and taking only ER positive tumors, the most relevant resistance generating mechanisms seems to be the alternative pro-growth signaling pathways up-regulation and the maintained ER triggered signaling irrespective of estrogenic deprivation following modulation of the pathway by other signal transducers. This phenomenon responds to the general rule of tumor resistance to oncologic treatments based on the plasticity of tumors and their capacity to adapt to a full range of adverse conditions based on the high number of pro-neoplasic mutation that originate cancer .

The availability of a growing body of signal transducers to combine with endocrine treatments has generated great expectations. Initial findings in the clinical settings are, nevertheless, modest. This fact highlights the need for a global understanding of cancer and for a finer taxonomy when designing clinical trials. The era of two-marker based taxonomy is over and a dynamic and more comprehensive approach of tumor classification stands as the cornerstone of therapeutic improvements.

References

Argiris, A., C. X. Wang, et al. (2004). "Synergistic interactions between tamoxifen and trastuzumab (Herceptin)." *Clin Cancer Res* 10(4): 1409-20.

Boccardo, F., A. Rubagotti, et al. (2001). "Sequential tamoxifen and aminoglutethimide versus tamoxifen alone in the adjuvant treatment of postmenopausal breast cancer patients: results of an Italian cooperative study." *J Clin Oncol* 19(22): 4209-15.

Boulay, A., J. Rudloff, et al. (2005). "Dual inhibition of mTOR and estrogen receptor signaling in vitro induces cell death in models of breast cancer." *Clin Cancer Res* 11(14): 5319-28.

Brueggemeier, R. W., J. C. Hackett, et al. (2005). "Aromatase inhibitors in the treatment of breast cancer." *Endocr Rev* 26(3): 331-45.

Bulun, S. E., S. Sebastian, et al. (2003). "The human CYP19 (aromatase P450) gene: update on physiologic roles and genomic organization of promoters." *J Steroid Biochem Mol Biol* 86(3-5): 219-24.

Bulun, S. E., K. Takayama, et al. (2004). "Organization of the human aromatase p450 (CYP19) gene." *Semin Reprod Med* 22(1): 5-9.

Bunone, G., P. A. Briand, et al. (1996). "Activation of the unliganded estrogen receptor by EGF involves the MAP kinase pathway and direct phosphorylation." *Embo J* 15(9): 2174-83.

Campbell, R. A., P. Bhat-Nakshatri, et al. (2001). "Phosphatidylinositol 3-kinase/AKT-mediated activation of estrogen receptor alpha: a new model for anti-estrogen resistance." *J Biol Chem* 276(13): 9817-24.

Chan, C. M., L. A. Martin, et al. (2002). "Molecular changes associated with the acquisition of oestrogen hypersensitivity in MCF-7 breast cancer cells on long-term oestrogen deprivation." *J Steroid Biochem Mol Biol* 81(4-5): 333-41.

Chen, S., J. Ye, et al. (2005). "Positive and negative transcriptional regulation of aromatase expression in human breast cancer tissue." *J Steroid Biochem Mol Biol* 95(1-5): 17-23.

Chen, S., D. Zhou, et al. (1999). "Breast tumor aromatase: functional role and transcriptional regulation." *Endocr Relat Cancer* 6(2): 149-56.

Chia, S., W. Gradishar, et al. (2008). "Double-blind, randomized placebo controlled trial of fulvestrant compared with exemestane after prior nonsteroidal aromatase inhibitor therapy in postmenopausal women with hormone receptor-positive, advanced breast cancer: results from EFECT." *J Clin Oncol* 26(10): 1664-70.

Cole, P. A. and C. H. Robinson (1990). "Mechanism and inhibition of cytochrome P-450 aromatase." *J Med Chem* 33(11): 2933-42.

Coleman, R. E., L. M. Banks, et al. (2007). "Skeletal effects of exemestane on bone-mineral density, bone biomarkers, and fracture incidence in postmenopausal women with early breast cancer participating in the Intergroup Exemestane Study (IES): a randomised controlled study." *Lancet Oncol* 8(2): 119-27.

Coombes, R. C., E. Hall, et al. (2004). "A randomized trial of exemestane after two to three years of tamoxifen therapy in postmenopausal women with primary breast cancer." *N Engl J Med* 350(11): 1081-92.

Coombes, R. C., L. S. Kilburn, et al. (2007). "Survival and safety of exemestane versus tamoxifen after 2-3 years' tamoxifen treatment (Intergroup Exemestane Study): a randomised controlled trial." *Lancet* 369(9561): 559-70.

di Salle, E., G. Ornati, et al. (1992). "Exemestane (FCE 24304), a new steroidal aromatase inhibitor." *J Steroid Biochem Mol Biol* 43(1-3): 137-43.

Dombernowsky, P., I. Smith, et al. (1998). "Letrozole, a new oral aromatase inhibitor for advanced breast cancer: double-blind randomized trial showing a dose effect and improved efficacy and tolerability compared with megestrol acetate." *J Clin Oncol* 16(2): 453-61.

Dowsett, M. (1996). "Endocrine resistance in advanced breast cancer." *Acta Oncol* 35 Suppl 5: 91-5.

Dressman, M. A., T. M. Walz, et al. (2001). "Genes that co-cluster with estrogen receptor alpha in microarray analysis of breast biopsies." *Pharmacogenomics J* 1(2): 135-41.

Ellis, C. A., M. D. Vos, et al. (2003). "Tamoxifen and the farnesyl transferase inhibitor FTI-277 synergize to inhibit growth in estrogen receptor-positive breast tumor cell lines." *Breast Cancer Res Treat* 78(1): 59-67.

Ellis, M. J., A. Coop, et al. (2001). "Letrozole is more effective neoadjuvant endocrine therapy than tamoxifen for ErbB-1- and/or ErbB-2-positive, estrogen receptor-positive primary breast cancer: evidence from a phase III randomized trial." *J Clin Oncol* 19(18): 3808-16.

Filardo, E. J., J. A. Quinn, et al. (2000). "Estrogen-induced activation of Erk-1 and Erk-2 requires the G protein-coupled receptor homolog, GPR30, and occurs via trans-activation of the epidermal growth factor receptor through release of HB-EGF." *Mol Endocrinol* 14(10): 1649-60.

Forbes, J. F., J. Cuzick, et al. (2008). "Effect of anastrozole and tamoxifen as adjuvant treatment for early-stage breast cancer: 100-month analysis of the ATAC trial." *Lancet Oncol* 9(1): 45-53.

Frasor, J., J. M. Danes, et al. (2003). "Profiling of estrogen up- and down-regulated gene expression in human breast cancer cells: insights into gene networks and pathways underlying estrogenic control of proliferation and cell phenotype." *Endocrinology* 144(10): 4562-74.

Frasor, J., F. Stossi, et al. (2004). "Selective estrogen receptor modulators: discrimination of agonistic versus antagonistic activities by gene expression profiling in breast cancer cells." *Cancer Res* 64(4): 1522-33.

Gee, J. M., M. E. Harper, et al. (2003). "The antiepidermal growth factor receptor agent gefitinib (ZD1839/Iressa) improves antihormone response and prevents development of resistance in breast cancer in vitro." *Endocrinology* 144(11): 5105-17.

Geisler, J., B. Haynes, et al. (2002). "Influence of letrozole and anastrozole on total body aromatization and plasma estrogen levels in postmenopausal breast cancer patients evaluated in a randomized, cross-over study." *J Clin Oncol* 20(3): 751-7.

Geisler, J., N. King, et al. (1998). "In vivo inhibition of aromatization by exemestane, a novel irreversible aromatase inhibitor, in postmenopausal breast cancer patients." *Clin Cancer Res* 4(9): 2089-93.

Geisler, J., N. King, et al. (1996). "Influence of anastrozole (Arimidex), a selective, non-steroidal aromatase inhibitor, on in vivo aromatisation and plasma oestrogen levels in postmenopausal women with breast cancer." *Br J Cancer* 74(8): 1286-91.

Geisler, J. and P. E. Lonning (2005). "Aromatase inhibition: translation into a successful therapeutic approach." *Clin Cancer Res* 11(8): 2809-21.

Goss, P. E., P. Hadji, et al. (2007). "Effects of steroidal and nonsteroidal aromatase inhibitors on markers of bone turnover in healthy postmenopausal women." *Breast Cancer Res* 9(4): R52.

Goss, P. E., J. N. Ingle, et al. (2005). "Randomized trial of letrozole following tamoxifen as extended adjuvant therapy in receptor-positive breast cancer: updated findings from NCIC CTG MA.17." *J Natl Cancer Inst* 97(17): 1262-71.

Goss, P. E., S. Qi, et al. (2007). "The effects of atamestane and toremifene alone and in combination compared with letrozole on bone, serum lipids and the uterus in an ovariectomized rat model." *Breast Cancer Res Treat* 103(3): 293-302.

Hong, Y., B. Yu, et al. (2007). "Molecular basis for the aromatization reaction and exemestane-mediated irreversible inhibition of human aromatase." *Mol Endocrinol* 21(2): 401-14.

Howell, A., J. Cuzick, et al. (2005). "Results of the ATAC (Arimidex, Tamoxifen, Alone or in Combination) trial after completion of 5 years' adjuvant treatment for breast cancer." *Lancet* 365(9453): 60-2.

Howell, A., C. K. Osborne, et al. (2000). "ICI 182,780 (Faslodex): development of a novel, "pure" antiestrogen." *Cancer* 89(4): 817-25.

Howell, A., J. F. Robertson, et al. (2004). "Comparison of fulvestrant versus tamoxifen for the treatment of advanced breast cancer in postmenopausal women previously untreated with endocrine therapy: a multinational, double-blind, randomized trial." *J Clin Oncol* 22(9): 1605-13.

Huang, E., S. H. Cheng, et al. (2003). "Gene expression predictors of breast cancer outcomes." *Lancet* 361(9369): 1590-6.

Jacob, J., K. S. Sebastian, et al. (2006). "Membrane estrogen receptors: genomic actions and post transcriptional regulation." *Mol Cell Endocrinol* 246(1-2): 34-41.

Jakesz, R., W. Jonat, et al. (2005). "Switching of postmenopausal women with endocrine-responsive early breast cancer to anastrozole after 2 years' adjuvant tamoxifen: combined results of ABCSG trial 8 and ARNO 95 trial." *Lancet* 366(9484): 455-62.

Jiang, Y., S. L. Harlocker, et al. (2002). "Discovery of differentially expressed genes in human breast cancer using subtracted cDNA libraries and cDNA microarrays." *Oncogene* 21(14): 2270-82.

Johnston, S. R. and M. Dowsett (2003). "Aromatase inhibitors for breast cancer: lessons from the laboratory." *Nat Rev Cancer* 3(11): 821-31.

Johnston, S. R., J. Head, et al. (2003). "Integration of signal transduction inhibitors with endocrine therapy: an approach to overcoming hormone resistance in breast cancer." *Clin Cancer Res* 9(1 Pt 2): 524S-32S.

Johnston, S. R., L. A. Martin, et al. (2005). "Aromatase inhibitors: combinations with fulvestrant or signal transduction inhibitors as a strategy to overcome endocrine resistance." *J Steroid Biochem Mol Biol* 95(1-5): 173-81.

Jones, A. L., T. J. Powles, et al. (1992). "Adjuvant aminoglutethimide for postmenopausal patients with primary breast cancer: analysis at 8 years." *J Clin Oncol* 10(10): 1547-52.

Jordan, V. C. (2003). "Targeting antihormone resistance in breast cancer: a simple solution." *Ann Oncol* 14(7): 969-70.

Jordan, V. C. (2004). "Selective estrogen receptor modulation: concept and consequences in cancer." *Cancer Cell* 5(3): 207-13.

Kahlert, S., S. Nuedling, et al. (2000). "Estrogen receptor alpha rapidly activates the IGF-1 receptor pathway." *J Biol Chem* 275(24): 18447-53.

Kao, Y. C., T. Okubo, et al. (1999). "Induction of aromatase expression by aminoglutethimide, an aromatase inhibitor that is used to treat breast cancer in postmenopausal women." *Anticancer Res* 19(3A): 2049-56.

Kaufmann, M., E. Bajetta, et al. (2000). "Exemestane is superior to megestrol acetate after tamoxifen failure in postmenopausal women with advanced breast cancer: results of a phase III randomized double-blind trial. The Exemestane Study Group." *J Clin Oncol* 18(7): 1399-411.

Kushner, P. J., D. A. Agard, et al. (2000). "Estrogen receptor pathways to AP-1." *J Steroid Biochem Mol Biol* 74(5): 311-7.

Lonning, P. E. (1990). "Aminoglutethimide enzyme induction: pharmacological and endocrinological implications." *Cancer Chemother Pharmacol* 26(4): 241-4.

Lonning, P. E. (1999). "Cross-resistance to different aromatase inhibitors in breast cancer treatment." *Endocr Relat Cancer* 6(2): 251-7.

Lonning, P. E. (2002). "Aromatase inhibitors and inactivators for breast cancer therapy." *Drugs Aging* 19(4): 277-98.

Lonning, P. E., J. Geisler, et al. (2005). "Effects of exemestane administered for 2 years versus placebo on bone mineral density, bone biomarkers, and plasma lipids in patients with surgically resected early breast cancer." *J Clin Oncol* 23(22): 5126-37.

Lonning, P. E. and S. Kvinnsland (1988). "Mechanisms of action of aminoglutethimide as endocrine therapy of breast cancer." *Drugs* 35(6): 685-710.

Ma, C. X., A. A. Adjei, et al. (2005). "Human aromatase: gene resequencing and functional genomics." *Cancer Res* 65(23): 11071-82.

Mackay, A., A. Urruticoechea, et al. (2007). "Molecular response to aromatase inhibitor treatment in primary breast cancer." *Breast Cancer Res* 9(3): R37.

Martin, L. A., I. Farmer, et al. (2003). "Enhanced estrogen receptor (ER) alpha, ERBB2, and MAPK signal transduction pathways operate during the adaptation of MCF-7 cells to long term estrogen deprivation." *J Biol Chem* 278(33): 30458-68.

Masamura, S., S. J. Santner, et al. (1995). "Estrogen deprivation causes estradiol hypersensitivity in human breast cancer cells." *J Clin Endocrinol Metab* 80(10): 2918-25.

Mauriac, L., A. Keshaviah, et al. (2007). "Predictors of early relapse in postmenopausal women with hormone receptor-positive breast cancer in the BIG 1-98 trial." *Ann Oncol* 18(5): 859-67.

Miki, Y., T. Suzuki, et al. (2007). "Effects of aromatase inhibitors on human osteoblast and osteoblast-like cells: a possible androgenic bone protective effects induced by exemestane." *Bone* 40(4): 876-87.

Miller, W. R., T. J. Anderson, et al. (2003). "An integrated view of aromatase and its inhibition." *J Steroid Biochem Mol Biol* 86(3-5): 413-21.

Miller, W. R. and J. O'Neill (1987). "The importance of local synthesis of estrogen within the breast." *Steroids* 50(4-6): 537-48.

Mouridsen, H., M. Gershanovich, et al. (2001). "Superior efficacy of letrozole versus tamoxifen as first-line therapy for postmenopausal women with advanced breast cancer: results of a phase III study of the International Letrozole Breast Cancer Group." *J Clin Oncol* 19(10): 2596-606.

Murray, R. and P. Pitt (1995). "Aromatase inhibition with 4-OHAndrostenedione after prior aromatase inhibition with aminoglutethimide in women with advanced breast cancer." *Breast Cancer Res Treat* 35(3): 249-53.

Normanno, N., M. Di Maio, et al. (2005). "Mechanisms of endocrine resistance and novel therapeutic strategies in breast cancer." *Endocr Relat Cancer* 12(4): 721-47.

Osborne, C. K. (1998). "Tamoxifen in the treatment of breast cancer." *N Engl J Med* 339(22): 1609-18.

Osborne, C. K., J. Pippen, et al. (2002). "Double-blind, randomized trial comparing the efficacy and tolerability of fulvestrant versus anastrozole in postmenopausal women with advanced breast cancer progressing on prior endocrine therapy: results of a North American trial." *J Clin Oncol* 20(16): 3386-95.

Osborne, C. K. and R. Schiff (2003). "Growth factor receptor cross-talk with estrogen receptor as a mechanism for tamoxifen resistance in breast cancer." *Breast* 12(6): 362-7.

Pasqualini, J. R., G. Chetrite, et al. (1996). "Concentrations of estrone, estradiol, and estrone sulfate and evaluation of sulfatase and aromatase activities in pre- and postmenopausal breast cancer patients." *J Clin Endocrinol Metab* 81(4): 1460-4.

Perou, C. M., S. S. Jeffrey, et al. (1999). "Distinctive gene expression patterns in human mammary epithelial cells and breast cancers." *Proc Natl Acad Sci U S A* 96(16): 9212-7.

Perou, C. M., T. Sorlie, et al. (2000). "Molecular portraits of human breast tumours." *Nature* 406(6797): 747-52.

Polychronis, A., H. D. Sinnett, et al. (2005). "Preoperative gefitinib versus gefitinib and anastrozole in postmenopausal patients with oestrogen-receptor positive and epidermal-growth-factor-receptor-positive primary breast cancer: a double-blind placebo-controlled phase II randomised trial." *Lancet Oncol* 6(6): 383-91.

Razandi, M., A. Pedram, et al. (2003). "Proximal events in signaling by plasma membrane estrogen receptors." *J Biol Chem* 278(4): 2701-12.

Robertson, J. F., C. K. Osborne, et al. (2003). "Fulvestrant versus anastrozole for the treatment of advanced breast carcinoma in postmenopausal women: a prospective combined analysis of two multicenter trials." *Cancer* 98(2): 229-38.

Santen, R. J., S. Santner, et al. (1978). "Aminoglutethimide inhibits extraglandular estrogen production in postmenopausal women with breast carcinoma." *J Clin Endocrinol Metab* 47(6): 1257-65.

Santen, R. J., R. X. Song, et al. (2005). "Long-term estradiol deprivation in breast cancer cells up-regulates growth factor signaling and enhances estrogen sensitivity." *Endocr Relat Cancer* 12 Suppl 1: S61-73.

Schiff, R., S. Massarweh, et al. (2003). "Breast cancer endocrine resistance: how growth factor signaling and estrogen receptor coregulators modulate response." *Clin Cancer Res* 9(1 Pt 2): 447S-54S.

Schiff, R., S. A. Massarweh, et al. (2004). "Cross-talk between estrogen receptor and growth factor pathways as a molecular target for overcoming endocrine resistance." *Clin Cancer Res* 10(1 Pt 2): 331S-6S.

Shiau, A. K., D. Barstad, et al. (1998). "The structural basis of estrogen receptor/coactivator recognition and the antagonism of this interaction by tamoxifen." *Cell* 95(7): 927-37.

Smith, I. E. (2003). "Letrozole versus tamoxifen in the treatment of advanced breast cancer and as neoadjuvant therapy." *J Steroid Biochem Mol Biol* 86(3-5): 289-93.

Smith, I. E., M. Dowsett, et al. (2005). "Neoadjuvant treatment of postmenopausal breast cancer with anastrozole, tamoxifen, or both in combination: the Immediate Preoperative Anastrozole, Tamoxifen, or Combined with Tamoxifen (IMPACT) multicenter double-blind randomized trial." *J Clin Oncol* 23(22): 5108-16.

Sorlie, T., C. M. Perou, et al. (2001). "Gene expression patterns of breast carcinomas distinguish tumor subclasses with clinical implications." *Proc Natl Acad Sci U S A* 98(19): 10869-74.

Sotiriou, C., T. J. Powles, et al. (2002). "Gene expression profiles derived from fine needle aspiration correlate with response to systemic chemotherapy in breast cancer." *Breast Cancer Res* 4(3): R3.

Sourdaine, P., M. G. Parker, et al. (1994). "Analysis of the aromatase cytochrome P450 gene in human breast cancers." *J Mol Endocrinol* 13(3): 331-7.

van 't Veer, L. J., H. Dai, et al. (2002). "Gene expression profiling predicts clinical outcome of breast cancer." *Nature* 415(6871): 530-6.

van de Vijver, M. J., Y. D. He, et al. (2002). "A gene-expression signature as a predictor of survival in breast cancer." *N Engl J Med* 347(25): 1999-2009.

Wang, X. and S. Chen (2006). "Aromatase destabilizer: novel action of exemestane, a food and drug administration-approved aromatase inhibitor." *Cancer Res* 66(21): 10281-6.

West, M., C. Blanchette, et al. (2001). "Predicting the clinical status of human breast cancer by using gene expression profiles." *Proc Natl Acad Sci U S A* 98(20): 11462-7.

Winer, E. P., C. Hudis, et al. (2005). "American Society of Clinical Oncology technology assessment on the use of aromatase inhibitors as adjuvant therapy for postmenopausal women with hormone receptor-positive breast cancer: status report 2004." *J Clin Oncol* 23(3): 619-29.

Zajchowski, D. A., M. F. Bartholdi, et al. (2001). "Identification of gene expression profiles that predict the aggressive behavior of breast cancer cells." *Cancer Res* 61(13): 5168-78.

In: Aromatase Inhibitors: Types, Mode of Action and Indications ISBN: 978-1-60741-711-8
Editor: Jean R. Lamonte

Chapter III

Aromatase Inhibitors

Mosammat Rashida Begum and Maruf Siddiqui
Dhaka Medical College and Hospital, Dhaka, Bangladesh

Abstract

Aromatase inhibitors (AIs) are anti-estrogenic agents, which exert their action by inhibiting aromatase enzyme. AIs are a newer class of drugs that inhibit the function of the aromatase enzyme. In postmenopausal women, AIs can suppress most of the peripheral aromatase activity, leading to profound estrogen deprivation. AIs are classified as first, second or third generation according to their sequential development. Aromatization is the final step in steroid biosynthesis and, therefore, aromatase is a suitable target for selective inhibition. Aromatase is expressed primarily in the ovary and also in central and peripheral tissues, fat, muscle, liver, and breast. Normal and malignant breast tissue contributes to the peripheral synthesis of estrogens. Studies have consistently shown that lifetime exposure to estrogens increase the risk of breast cancer. The degree of risk is increased by persistently elevated blood concentrations of estrogen. AI can effectively reduce the estrogen level hence improves breast cancer including advanced stage.

Aromatase enzyme is demonstrated locally in endometriotic implants, which indicates local production of estrogen by these implants. Thus, extraovarian aromatse is likely to cause a resistance to GnRHa treatment via persistence of local and peripheral estrogen production during such treatment. In this situation aromatase inhibitors are effective to reduce the synthesis of both ovarian and extraovarian estrogen.

Anti-estrogens have lot of potential to be used as oral ovulation inducing agent. In the late 1990s, aromatase inhibitors, including letrozole, began to be used to induce ovulation by being administered in the early part of the menstrual cycle. Estrogen production from all sources is blocked by inhibiting aromatization, releasing the hypothalamic-pituitary axis from estrogenic negative feedback and resulting in increased gonadotropin secretion and ovarian follicular stimulation.

There may be potential new applications for aromatase inhibitors including improving implantation and prevention of ovarian hyperstimulation syndrome during assisted reproductive technology by reducing supraphysiologic level of estrogen.

Regarding safety concern different studies showed that there were no increased rates of major and minor malformations beyond what would be expected in the general population (1-3%).

It is not far away that aromatase inhibitors will possibly emerge as first line ovulation inducing drug for anovulatory or ovulatory infertility following further large-scale, multicentric, randomized controlled research and clinical trials.

Introduction

Aromatase enzyme is the product of CYP19 gene, which is highly expressed in human placenta and in granulose cells of ovarian follicles, where its expression depends on cyclical gonadotropin stimulation. Aromatase is also present, at lower levels, in several non-glandular tissues, including skin and adipose tissue, liver, muscle, brain, normal breast and breast cancer tissue [1]. The aromatase enzyme is responsible for the conversion of androstenedione and testosterone to the estrogens, estrone (E1) and estradiol (E2), respectively. In postmenopausal women, this conversion occurs primarily in peripheral tissues while estrogen production in premenopausal women is primarily in the ovary.

Aromatase inhibitors (AIs) are anti-estrogenic agents, which exert their action by inhibiting aromatase enzyme. It was first developed in 1960's and over last four decades more potent and selective AIs with less toxic effects have evolved. AIs are a newer class of drugs that inhibit the function of the aromatase enzyme. In postmenopausal women, AIs can suppress most of the peripheral aromatase activity, leading to profound estrogen deprivation.

Types of Aromatase Inhibitors

AIs are classified as first, second or third generation according to their sequential development. In addition, they are further classified as type 1 (steroidal aromatase inactivator) or type 2 (nonsteroidal aromatase inhibitor) according to their mechanism of action. Type 1 inhibitors are steroidal analogs of androstenedione. Androstenedione analogs bind to the same site on the aromatase molecule, but unlike androstenedione, they bind reversibly because of their conversion to reactive intermediates by aromatase. Thus, they are commonly known as aromatase inactivators. Type 2 inhibitors are nonsteroidal and bind reversibly to the haem group of cytochrome p450 enzyme by way of a basic nitrogen atom [1].

Aminoglutethimide, a first-generation type 2 AI, originally developed as an anticonvulsant was subsequently found to be a weak AI. It induced medical adrenalectomy by inhibiting many other enzymes involved in steroid biosynthesis. Because of significant toxicities, its use has declined considerably.

Second-generation AIs include *formestane,* a type 1 steroidal inactivator, *fadrozole*, a type 2 imidazole; and *rogletimide,* a type 2 inhibitor. Formestane was the first AI widely used for breast cancers. However, with the advent of newer AIs, use of this group has declined considerably.

Third generation AIs which are in current clinical use include the type 1 steroidal agent *exemestane* and the type 2 nonsteroidal imidazoles, *anastrazole* and *letrozole*.

Mode of Action

Aromatase is an enzyme (cytochrome P-450 [CYP] 19), which catalyzes the rate-limiting step (conversion of androgens to estrogens) in estrogen biosynthesis [2-4]. Aromatization is the final step in steroid biosynthesis [5] and, therefore, aromatase is a suitable target for selective inhibition [6,7]. Aromatase is expressed primarily in the ovary and also in central and peripheral tissues, fat, muscle, liver, and breast [8,9]. With increasing age, as ovarian estrogen production declines [10] the contribution of peripheral production of estrogens increases [11] and in postmenopausal women, peripheral aromatization of androstenedione produced by the adrenal gland becomes the main source of endogenous estrogens [12-16]. Normal and malignant breast tissue contributes to the peripheral synthesis of estrogens [17-21].

Modern third-generation AIs effectively block the production of estrogen without exerting effects on other steroidogenic pathways [22]. The first-generation inhibitor aminoglutethimide lacked selectivity for aromatase [23] and inhibited biosynthesis of cortisol, aldosterone, and thyroid hormone [24] as well as aromatase; moreover, aminoglutethimide was also found to induce hepatic enzymes [25,26]. Second-generation AIs included the nonsteroidal inhibitor fadrozole and the steroidal inhibitor formestane (4-hydroxyandrostenedione). Fadrozole was superior to aminoglutethimide in terms of potency, selectivity, and safety [27] but its selectivity was not complete and clinical trials suggested that it was no more effective than tamoxifen [28,29].

The third-generation AI letrozole (Femara®) is the most highly potent and totally selective AI [30], which inhibits aromatase only without effecting biosynthesis of other steroids.

Aromatase inhibitors block the action of aromatase and prevents conversion of androgen to estrogen. As a result estrogen level in systemic circulation falls tremendously. At a dose of 1-5 mg/day they inhibit estrogen levels by 95% [31]. This level of inhibition results in almost undetectable level of estradiol in sensitive immunoassay. Thus it regress the growth of estrogen dependant target tissue. Lower estrogen inhibits negative feedback mechanism at pituitary and mediobasal hypothalamus, thus stimulates the secretion of FSH and LH in reproductive aged women.

Letrozole is completely absorbed after oral administration with a mean half life of 42 hours. AIs are cleared from systemic circulation mainly by the liver [32]. Gastrointestinal disturbances, asthenia, hot flushes, headache and back pain are main adverse effects of the drug [33-36].

Indications of Use

The third generation aromatase inhibitors are FDA approved drug to treat postmenopausal breast cancer. However, in the field of gynaecological practice, many other potential indications of this drug have gradually evolved. Recent evidence of different researches is suggestive that aromatase inhibitors can potentially be used to treat endometriosis, uterine leiomyoma and to induce or augment ovulation.

Use in Postmenopausal Breast Cancer

Studies have consistently shown that lifetime exposure to estrogens increase the risk of breast cancer [37]. The degree of risk is increased by persistently elevated blood concentrations of estrogen [38]. Estrogen is thought to contribute to the initiation, promotion and progression of breast cancer via two complementary mechanisms [37], the carcinogenic effects of estrogen metabolites, notably hydroxyl metabolites [39-41], and stimulation of ER signaling pathways [42-53]. Hormone receptor (HR)-positive tumors are defined as those with ER or progesterone receptor (PgR) expression detectable above a pre-set limit [54]. Approximately two thirds of breast cancer patients have HR+ tumors [55] and are candidates for treatment strategies designed to counteract the growth effects of estrogen.

Exemestane, used in postmenopausal breast cancer patients, has less androgenic activity than formestane but has a similar toxicity profile. Exemstane has been tested in a randomized clinical trial in postmenopausal women with estrogen-receptor positive breast cancer. Women who had completed 2 to 3 years of adjuvant tamoxifen were randomized to complete a total of 5 years of adjuvant treatment with tamoxifen or exemstane. Though overall survival was not significantly different in two groups, exemstane demonstrated an absolute benefit in terms of disease free interval [56].

Anastrazole and letrozole both bind competitively and specifically to the haem of CYP19. Used for the treatment of estrogen receptor positive postmenopausal breast cancer patients, they both showed a statistically significant survival period over tamoxifen.

Anastrazole, 1 mg oral daily dose, has shown efficacy in the treatment of postmenopausal women with early stage or advanced, hormone-receptor positive breast cancer. In early stage breast cancer, anastrazole was significantly more effective than tamoxifen in terms of time to tumor recurrence and odds of a primary contralateral tumor [57].

Letrozole, 2.5 mg administered orally once daily has shown efficacy in the treatment of postmenopausal women with early stage or advanced, hormone receptor positive breast cancer. In early-stage disease, extending adjuvant endocrine therapy with letrozole (beyond the standard 5-year period of tamoxifen) improved disease-free survival compared to placebo [58] and improved overall survival in the subset of patients with positive axillary nodes [59].

In advanced breast cancer, letrozole is superior to tamoxifen as first-line treatment. Time to disease progression is significantly higher and objective response rate is significantly greater with letrozole but median overall survival is similar between the groups. As second-line therapy of advanced breast cancer that has progressed to anti-estrogen therapy, letrozole shows efficacy equivalent to that of anastrazole [60].

Letrozole has no significant active metabolites. Letrozole generally is well tolerated; the most common treatment-related adverse effects are hot flushes, nausea, and hair thinning. In the trial of extended adjuvant therapy, adverse events reported more frequently with letrozole than placebo were hot flushes, arthralgia, myalgia and arthritis, but cessation of letrozole was no more frequent than placebo in this double-blind trial. A greater number of new diagnosis of osteoporosis occurred among women receiving letrozole, but the long term effects on bone mineral density or lipid metabolism are yet to be determined.

Use in Endometriosis and Fibroid Uterus

Aromatase enzyme is demonstrated locally in endometriotic implants, which indicates local production of estrogen by these implants. As a result, progression of endometriosis continues even during treatment with GnRH analogues, which only inhibit ovarian production of estrogen [61]. Besides skin and adipose tissue also contain aromatase and continue to generate significant circulating levels of E2 during GnRH analogue treatment [62]. Thus, extraovarian aromatse is likely to cause a resistance to GnRHa treatment via persistence of local and peripheral estrogen production during such treatment. In this situation aromatase inhibitors are effective to reduce the synthesis of both ovarian and extraovarian estrogen, thereby regress the development of emdometriotic implants. By the same mechanism aromatase inhibitors can reduce the size of the fibroids.

Use in Infertility

Anti-estrogens have lot of potential to be used as oral ovulation inducing agent. Hence, aromatase inhibitors are increasingly used for ovulation induction now a days as an unapproved indication. Traditionally, anti-estrogens such as the clomiphene citrate have been the first choice for induction of ovulation in WHO type II anovulatory infertility where intact Hypothalamo-pituitary axis is present. However, 20-25% women are resistant to clomiphene citrate [63]. In addition, clinical data reveal a discrepancy between ovulation (60-80%) and conception rates (20-40%) between clomiphene citrate (CC) treatment, and a higher than expected incidence of miscarriage in conception cycles. These observations have been attributed to the peripheral anti-estrogenic mechanism of action of CC that involves long lasting estrogen receptor depletion in the endometrium and cervix [63]. It also appears that CC accumulates in the body because of long half-life leading to negative effect on endometrial development and quality and quantity on cervical mucous. High concentration of CC and its isomers may have some harmful effects on embryo development and oocyte quality. Diminished uterine blood flow during early luteal phase and pre-implantation stage is another explanation of poor outcome. Attempt to overcome these drawbacks of CC with either addition of estrogens along with CC or replacing by other SERM like tamoxiphene did not improve pregnancy outcome or has failed to show any convincing improvement in other parameters.

In the late 1990s, aromatase inhibitors, including letrozole, began to be used to induce ovulation by being administered in the early part of the menstrual cycle. Estrogen production from all sources is blocked by inhibiting aromatization, releasing the hypothalamic-pituitary axis from estrogenic negative feedback and resulting in increased gonadotropin secretion and ovarian follicular stimulation. In the ovary, aromatase inhibitors increase follicular sensitivity to FSH, as there is an accumulation of intraovarian androgens.

At the level of the endometrium, estrogen receptors may be upregulated, resulting in rapid endometrial growth once estrogen secretion is restored. In its ability to induce ovulation, letrozole compares favourably to clomiphene citrate, which has been the first-line treatment for ovulatory disorders for more than 40 years, and it has emerged as an alternative to CC for ovulation induction. The clinical outcome of early pregnancies achieved through the use of letrozole for ovulation induction or controlled ovarian hyperstimulation for intrauterine insemination was reported in 2005 in a cohort study. The outcomes of pregnancies achieved through letrozole and other ovarian stimulation regimens (CC) were compared with a control group composed of women who had conceived spontaneously.

Infertility specialists have been using this drug for ovulation induction since 2001.Over 200 trials using letrozole are already done in various countries (eg. Canada, USA, UK, Greece, Netherlands, Germany, Korea, Belgium, Peru, Thailand, Mexico, Bangladesh, Taiwan, Turkey, India etc.). letrozole has been extensively studied at the University of Toronto as a fertility drug with satisfying results [64]. Their studies found that the drug appears to be completely eliminated from the body within a few days of the last tablet. Because of the much shorter life span (42 hrs) and absence of estrogen receptor binding, it possibly avoids the anti-estrogenic effects associated with CC.

Evidences of Clinical Trial

Over the last few years, several published studies, both controlled and noncontrolled, compared clomiphene citrate and letrozole, either alone or in combination with gonadotropins, for ovulation induction or augmentation [65].These studies found letrozole as effective as clomiphene in inducing ovulation, with the major advantage of absence of any anti-estrogenic adverse effects.

Patients treated with letrozole demonstrated low estrogen production per developing follicle resulting in more physiological estrogen levels around the time of ovulation and good pregnancy rates with a lower incidence of multiple pregnancies than with clomiphene [65]. When combined with gonadotropins, letrozole reduces the dose of gonadotropins required for optimal follicular recruitment and improve the response to gonadotropin stimulation in poor responders [66]. However, a recent study involving 438 PCOS patients did not show any advantage of letrozole over CC as a first line treatment for induction of ovulation in PCOS patients [67]. A group of Turkish researchers concluded that letrozole and CC have comparable effectiveness in ovulatory patients with unexplained infertility [68]. A small study involving 22 patients from the University of Toronto showed the supremacy of letrozole in ovulation induction in patients with an inadequate response to clomiphene citrate [69]. This hypothesis is supported by another study in a tertiary referral infertility clinic at

Dhaka involving 35 infertile patients poor responsive to CC, which showed a high success rate (90%) for follicular development using letrozole [70]. They also reported that ovulation rate was 1.5 times (62.5%) higher with letrozole than with CC (37.5%) in PCOS patients [71]. Another survey also reported a successful induction (62%) in PCOS patients using letrozole previously treated with clomiphene citrate [72]. Researchers from a university hospital in Greece concluded that ovarian induction with letrozole is associated with acceptable success rate compared with gonadotropin with significant less cost, risks and patient inconvenience [73]. In patients who failed to conceive with CC, gonadotropins have a higher pregnancy rate for ovulation induction than letrozole [74]. However, pregnancy rates were high enough with letrozole to justify its use. Compared to CC in gonadotropin combined Intrauterine Insemination (IUI) cycles, number of mature follicles and serum estradiol levels on the day of hCG were significantly lower in the letrozole group. No significant differences were found in endometrial thickness measured on the day of hCG but the pregnancy rate was higher in letrozole group [75]. Comparing the endocrinological environment, significantly lower estradiol concentrations and fewer follicles were observed in cycles stimulated with 2.5 mg letrozole compared to 100 mg of CC from day 3 to day 7 of the cycle [76]. However, the administration of 50 mg CC was found superior to 2.5 mg letrozole for superovulation induction in Asian women with normal ovulation [77]. Similar to CC, aromatase inhibition by letrozole reduces FSH dose required for controlled ovarian hyperstimulation (COH) without the undesirable anti-estrogenic effects sometimes seen with CC. Letrozole appears to constitute a good alternative to CC in patients with unexplained infertility undergoing gonadotropin stimulated COH cycles combined with IUI therapy [78]. It also seems to be an acceptable alternative to CC as an ovulation inducing agent in patients with PCOS [79]. The third generation aromatase inhibitor letrozole is as effective as or superior to CC in ovulation induction and in superovulation [80].

Unlike CC, which blocks and depletes estrogen receptors, aromatase inhibitors have no effect on estrogen receptors. Aromatase inhibitors should, therefore, not have any deleterious effect on cervical mucus or endometrium like CC which interferes with the attainment of a pregnancy and is probably the main reason for the gap between ovulation and pregnancy rates. This can theoretically be avoided when aromatase inhibitors are used for the same purpose. With aromatase inhibitors, estrogen production is eventually advanced by the induced FSH discharge but, in contrast to the use of CC, the hypothalamus is able to respond to the estrogen feedback with a negative feedback mechanism. This will modulate an overzealous discharge of FSH which in turn is more likely to result in a monofollicular ovulation with moderate estrogen concentrations. This confers an additional possible advantage to aromatase inhibitors as the prevalence of multiple pregnancies could therefore be expected to be less than that with the use of CC for ovulation induction.

Aromatase inhibitors result in an accumulation of androgens whose conversion to estrogens is being blocked. This would be an unwanted by-product, especially for women with PCOS who already have an excessive production of androgens. But androgens may have a stimulatory role in early follicular growth by augmenting follicular FSH receptor expression and therefore amplifying FSH effect [81-84]. This may explain the relative success of combined letrozole and FSH for ovarian stimulation in improving the response to FSH, reported by Mitwally and Casper [85] –and Healey *et al* [86]. A lower FSH dose and a significantly higher number of mature follicles were achieved with the combined treatment [85,86]. The addition of letrozole to gonadotropin treatment again decreased the dose of gonadotropins and increased the number of pre-ovulatory follicles. Prospective, randomized trials are needed to verify these interesting findings. Letrozole is yet to undergo further large, robust clinical trials in order to fulfill this promise as a feasible alternative to CC in this role, but some preliminary trials have shown at least equality and some superiority.

Future Avenues of Aromatase Inhibitor Use

There may be potential new applications for aromatase inhibitors including improving implantation and prevention of ovarian hyperstimulation syndrome during assisted reproductive technology. Low implantation and pregnancy rates relative to fertilization and cleavage rates are thought to be related in part to ovarian stimulation [87]. Supraphysiologic levels of estrogen attained during ovarian stimulation may explain some of the adverse effects of ovarian stimulation on the outcome of infertility treatment including the lower pregnancy rates than expected [88-90]. Different mechanisms have been postulated to explain the adverse effect of supraphysiologic levels of estrogen on reproductive outcome including deleterious effects on the endometrium [91-93] the embryo [88, 94-97] the coagulation system [98,99] and oviduct [100]. Reducing estrogen synthesis by aromatase inhibitors may be a method of achieving amelioration of the deleterious effects of the supraphysiologic levels of estrogen during ovarian stimulation. Ovarian hyperstimualtion syndrome is associated with high levels of estrogen production. Reducing the synthesis of estrogen by aromatase inhibitors and thereby reducing the development of ovarian hyperstimulation syndrome may be another potential use of this drug.

Safety of the Drug

In 2005, some concern was raised as a result of an abstract of a study that compared 150 babies born to women who had used letrozole with 36,005 babies born to low-risk pregnant women [101]. The results of this study, which had several gross methodological drawbacks, suggested that letrozole might increase the risk of cardiac and bone anomalies [101]. Since that initial report, however, several full-length studies have not shown an increased risk for congenital malformations following the use of letrozole to induce ovulation. A study published in 2006 examined 514 babies born to mother who had used letrozole to conceive, and compared them with 397 babies born to mothers who had conceived using clomiphene citrate [102]. There were no increased rates of major and minor malformations beyond what would be expected in the general population (1-3%). Additionally, the number of cardiac anomalies in the letrozole group (0.2%) was slightly lower than that of the general population [102].

Conclusions

In clomiphene citrate failures either as a result of persistent anovulation or of anti-estrogenic effects on the endometrium or cervical mucus, gonadotropin injections are generally used, which needs extensive monitoring to avoid hyperstimulation. In these cases aromatase inhibitors may be the next alternative to clomiphene citrate instead of potentially risky gonadotropins. It is not far away that aromatase inhibitors will possibly emerge as first line ovulation inducing drug for anovulatory or ovulatory infertility following further large-scale, multicentric, randomized controlled research and clinical trials.

References

[1] Smith IE and Dowsett M. Aromatase inhibitors in breast cancer. *N Engl J Med.* 2003,348:2431-2442.

[2] Ryan KJ. Biological aromatization of steroids. *J Biol Chem* 1959;234:268–272.

[3] Meigs RA, Ryan KJ. Cytochrome P-450 and steroid biosynthesis in the human placenta. *Biochim Biophys Acta* 1968;165:476–482 .

[4] Fishman J, Raju MS. Mechanism of estrogen biosynthesis. Stereochemistry of C-1 hydrogen elimination in the aromatization of 2 beta-hydroxy-19-oxoandrostenedione. *J Biol Chem* 1981;256:4472–4477.

[5] Lønning PE. Aromatase inhibitors in breast cancer. *Endocr Relat Cancer* 2004;11:179–189.

[6] Brodie AM, Schwarzel WC, Shaikh AA, Brodie HJ. The effect of an aromatase inhibitor, 4-hydroxy-4-androstene-3,17-dione, on estrogen-dependent processes in reproduction and breast cancer. *Endocrinology* 1977;100:1684–1695.

[7] Brodie AM, Brodie HJ, Garrett WM, Hendrickson JR, Marsh DA, Tsai-Morris CH. Effect of an aromatase inhibitor, 1,4,6-androstatriene-3,17-dione, on 7,12-dimethylbenz[a]anthracene-induced mammary tumors in the rat and its mechanism of action in vivo. *Biochem Pharmacol* 1982;31:2017–2023.

[8] Longcope C, Pratt JH, Schneider SH, Fineberg SE. Aromatization of androgens by muscle and adipose tissue in vivo. *J Clin Endocrinol Metab* 1978;46:146–152.

[9] Miller WR. Aromatase activity in breast tissue. *J Steroid Biochem Mol Biol* 1991;39:783–790.

[10] Couzinet B, Meduri G, Lecce MG, Young J, Brailly S, Loosfelt H, Milgrom E, Schaison G. The postmenopausal ovary is not a major androgen-producing gland. *J Clin Endocrinol Metab* 2001;86:5060–5066.

[11] Hemsell DL, Grodin JM, Brenner PF, Siiteri PK, MacDonald PC. Plasma precursors of estrogen. II. Correlation of the extent of conversion of plasma androstenedione to estrone with age. *J Clin Endocrinol Metab* 1974;38:476–479.

[12] Schweikert HU, Milewich L, Wilson JD. Aromatization of androstenedione by cultured human fibroblasts. *J Clin Endocrinol Metab* 1976;43:785–795.

[13] Smuk M, Schwers J. Aromatization of androstenedione by human adult liver in vitro. *J Clin Endocrinol Metab* 1977;45:1009–1012.

[14] Perel E, Killinger DW. The interconversion and aromatization of androgens by human adipose tissue. *J Steroid Biochem* 1979;10:623–627.

[15] Frisch RE, Canick JA, Tulchinsky D. Human fatty marrow aromatizes androgen to estrogen. *J Clin Endocrinol Metab* 1980;51:394–396.

[16] Matsumine H, Hirato K, Yanaihara T, Tamada T, Yoshida M. Aromatization by skeletal muscle. *J Clin Endocrinol Metab* 1986;63:717–720.

[17] Miller WR, Forrest AP. Oestradiol synthesis by a human breast carcinoma. *Lancet* 1974;2:866–868.

[18] Perel E, Wilkins D, Killinger DW. The conversion of androstenedione to estrone, estradiol, and testosterone in breast tissue. *J Steroid Biochem* 1980;13:89–94.

[19] Reed MJ, Owen AM, Lai LC, Coldham NG, Ghilchik MW, Shaikh NA, James VH. In situ estrone synthesis in normal breast and breast tumor tissues: effect of treatment with 4-hydroxyandrostenedione. *Int J Cancer* 1989;44:233–237.

[20] Miller WR, Anderson TJ, Jack WJ. Relationship between tumor aromatase activity, tumor characteristics and response to therapy. *J Steroid Biochem Mol Biol* 1990;37:1055–1059.

[21] Bulun SE, Price TM, Aitken J, Mahendroo MS, Simpson ER. A link between breast cancer and local estrogen biosynthesis suggested by quantification of breast adipose tissue aromatase cytochrome P450 transcripts using competitive polymerase chain reaction after reverse transcription. *J Clin Endocrinol Metab* 1993;77:1622–1628. Erratum in: *J Clin Endocrinol Metab* (1994) 78:494.

[22] Swain SM. Aromatase inhibitors—a triumph of translational oncology. *N Engl J Med* 2005;353:2807–2809.

[23] Hausler A, Schenkel L, Krahenbuhl C, Monnet G, Bhatnagar AS. An in vitro method to determine the selective inhibition of estrogen biosynthesis by aromatase inhibitors. *J Steroid Biochem* 1989;33:125–131.

[24] Pittman JA, Brown RW. Antithyroid and antiadrenocortical activity of aminoglutethimide. *J Clin Endocrinol Metab* 1966;26:1014–1016.

[25] Murray M, Cantrill E, Farrell GC. Induction of cytochrome P450 2B1 in rat liver by the aromatase inhibitor aminoglutethimide. *J Pharmacol Exp Ther* 1993;265:477–481.

[26] Santen RJ, Harvey HA. Use of aromatase inhibitors in breast carcinoma. *Endocr Relat Cancer* 1999;6:75–92.

[27] Santen RJ, Demers LM, Lynch J, Harvey H, Lipton A, Mulagha M, Hanagan J, Garber JE, Henderson IC, Navari RM, Miller AA. Specificity of low dose fadrozole hydrochloride (CGS 16949A) as an aromatase inhibitor. *J Clin Endocrinol Metab* 1991;73:99–106.

[28] Falkson CI, Falkson HC. A randomised study of CGS 16949A (fadrozole) versus tamoxifen in previously untreated postmenopausal patients with metastatic breast cancer. *Ann Oncol* 1996;7:465–469.

[29] Thürlimann B, Beretta K, Bacchi M, Castiglione-Gertsch M, Goldhirsch A, Jungi WF, Cavalli F, Senn H-J, Fey M, Löhnert T. First-line fadrozole HCI (CGS 16949A) versus tamoxifen in postmenopausal women with advanced breast cancer. Prospective randomised trial of the Swiss Group for Clinical Cancer Research SAKK 20/88. *Ann Oncol* 1996;7:471–479.

[30] Bhatnagar AS, Häusler A, Schieweck K, Lang M, Bowman R. Highly selective inhibition of estrogen biosynthesis by CGS 20267, a new non-steroidal aromatase inhibitor. *J Steroid Biochem Mol Biol* 1990;37:1021–1027.

[31] Iveson TJ, Smith IE, Ahern J, Smithers DA, Trunet PF, Dowsett M. Phase I study of the oral nonsteroidal aromatase inhibitor CGS 20267 in postmenopausal patients with advanced breast cancer. *Cancer Res* 1993;53:266–270.

[32] Okman KT, Kucuk M, Aitaner S. Comparison of the effects of letrozole and clomiphene citrate on ovarian follicles, endometrium and hormone levels in the rat. *Fertil Steril* 2003;80:1330-1332.

[33] BuzdarA, Jonat W, Howel A, et al. Anstrozole, a potent and selective aromatase inhibitor versus megestrol acetate in postmenopausal women with advance breast cancer: result of overview analysis of teo phase ill trials. *J Clin Oncol* 1996;14:2000-11.

[34] Marry M, Gershanovich M, Campos B et al. AIs a new potent, selective aromatase inhibitor superior to aminogluthethimide ((AG) in postmenopausal women with advanced breast cancer previously treated with antioestrogens. *Proc Am Soc Clin Oncol* 1977;16:156-61.

[35] Bergh J, Bonneterre J, Houston SJ, et al. Vorozole (Rivizor) versus aminoglutethimide (AG) in the treatment of postmenopausal breast cancer relapsing after tamoxiphen [abstract] *Proc Am Soc Clin Oncol* 1997;16:156.

[36] Dowsett M. Biological background to aromatase inhibition. *The breast* 1996;5:196-201.

[37] Yager JD, Davidson NE. Estrogen carcinogenesis in breast cancer. *N Engl J Med* 2006;354:270–282.

[38] Key TJ, Appleby PN, Reeves GK, Roddam A, Dorgan JF, Longcope C, Stanczyk FZ, Stephenson HE Jr, Falk RT, Miller R, Schatzkin A, Allen DS, Fentiman IS, Key TJ, Wang DY, Dowsett M, Thomas HV, Hankinson SE, Toniolo P, Akhmedkhanov A, Koenig K, Shore RE, Zeleniuch-Jacquotte A, Berrino F, Muti P, Micheli A, Krogh V, Sieri S, Pala V, Venturelli E, Secreto G, Barrett-Connor E, Laughlin GA, Kabuto M, Akiba S, Stevens RG, Neriishi K, Land CE, Cauley JA, Kuller LH, Cummings SR, Helzlsouer KJ, Alberg AJ, Bush TL, Comstock GW, Gordon GB, Miller SR, Longcope C; Endogenous Hormones Breast Cancer Collaborative Group. Body mass index, serum sex hormones, and breast cancer risk in postmenopausal women. *J Natl Cancer Inst* 2003;95:1218–1226.

[39] Yue W, Santen RJ, Wang JP, Li Y, Verderame MF, Bocchinfuso WP, Korach KS, Devanesan P, Todorovic R, Rogan EG, Cavalieri EL (2003) Genotoxic metabolites of estradiol in breast: potential mechanism of estradiol induced carcinogenesis. *J Steroid Biochem Mol Biol*2003; 86:477–486.

[40] Devanesan P, Santen RJ, Bocchinfuso WP, Korach KS, Rogan EG, Cavalieri E. Catechol estrogen metabolites and conjugates in mammary tumors and hyperplastic tissue from estrogen receptor-alpha knock-out (ERKO)/Wnt-1 mice: implications for initiation of mammary tumors. *Carcinogenesis* 2001;22:1573–1576.

[41] Cavalieri EL, Stack DE, Devanesan PD, Todorovic R, Dwivedy I, Higginbotham S, Johansson SL, Patil KD, Gross ML, Gooden JK, Ramanathan R, Cerny RL, Rogan

EG. Molecular origin of cancer: catechol estrogen-3,4-quinones as endogenous tumor initiators. *Proc Natl Acad Sci* USA 1997;94:10937–10942.

[42] Santen RJ, Lobenhofer EK, Afshari CA, Bao Y, Song RX (2005) Adaptation of estrogen-regulated genes in long-term estradiol deprived MCF-7 breast cancer cells. *Breast Cancer Res Treat*2005;94:213–223.

[43] Webb P, Nguyen P, Valentine C, Lopez GN, Kwok GR, McInerney E, Katzenellenbogen BS, Enmark E, Gustafsson JA, Nilsson S, Kushner PJ. The estrogen receptor enhances AP-1 activity by two distinct mechanisms with different requirements for receptor transactivation functions. *Mol Endocrinol.*1999;13:1672–1685.

[44] Sabbah M, Courilleau D, Mester J, Redeuilh G. Estrogen induction of the cyclin D1 promoter: involvement of a cAMP response-like element. *Proc Natl Acad Sci* USA 1999; 96:11217–11222.

[45] Levin ER. Bidirectional signaling between the estrogen receptor and the epidermal growth factor receptor. *Mol Endocrinol* 2003;17:309–317.

[46] Stoica GE, Franke TF, Moroni M, Mueller S, Morgan E, Iann MC, Winder AD, Reiter R, Wellstein A, Martin MB, Stoica A. Effect of estradiol on estrogen receptor-alpha gene expression and activity can be modulated by the ErbB2/PI 3-K/Akt pathway. *Oncogene* 2003;22:7998–8011.

[47] Sekeris CE. The mitochondrial genome: a possible primary site of action of steroid hormones. *In vivo* 1990;4:317–320.

[48] Liao JK (2003) Cross-coupling between the oestrogen receptor and phosphoinositide 3-kinase. *Biochem Soc Trans* 2003;31:66–70.

[49] Aronica SM, Kraus WL, Katzenellenbogen BS. Estrogen action via the cAMP signalling pathway: stimulation of adenylate cyclase and cAMP-regulated gene transcription. *Proc Natl Acad Sci* USA 1994;91:8517–8521.

[50] Migliaccio A, Di Domenico M, Castoria G, de Falco A, Bontempo P, Nola E, Auricchio F. Tyrosine kinase/p21ras/MAP-kinase pathway activation by estradiol-receptor complex in MCF-7 cells. *EMBO J* 1996;15:1292–1300.

[51] Song RX-D, McPherson RA, Adam L, Bao Y, Shupnik M, Kumar R, Santen RJ. Linkage of rapid estrogen action to MAPK activation by ERα-Shc association and Shc pathway activation. *Mol Endocrinol* 2002;16:116–127.

[52] Razandi M, Pedram A, Greene GL, Levin ER. Cell membrane and nuclear estrogen receptors (ERs) originate from a single transcript: studies of ERalpha and ERbeta expressed in Chinese hamster ovary cells. *Mol Endocrinol* 1999;13:307–319.

[53] Razandi M, Pedram A, Levin ER (2000) Plasma membrane estrogen receptors signal to antiapoptosis in breast cancer. *Mol Endocrinol* 2000;14:1434–1447.

[54] Goldhirsch A, Glick JH, Gelber RD, Coates AS, Thurlimann B, Senn HJ; Panel members. Meeting highlights: international expert consensus on the primary therapy of early breast cancer 2005. *Ann Oncol* 2005;16:1569–1583.

[55] Huang WY, Newman B, Millikan RC, Schell MJ, Hulka BS, Moorman PG (2000) Hormone-related factors and risk of breast cancer in relation to estrogen receptor and progesterone receptor status. *Am J Epidemiol*2000; 151:703–714.

[56] Coombes RC, Hall E, Gibson LJ et al. A randomized trial of exemestane after two to three years of tamoxifen therapy in postmenopausal women with primary breast cancer. *N Engl J Med.* 2004, 350:1081-1092. Erratum in: *N Engl J Med.* 2004,351:2461.
[57] Baum M, Budzar AU, Cuzick J et al. Anastrazole alone or in combination with tamoxifen versus tamoxifen alone for adjuvant treatment of postmenopausal women with early breast cancer: first results of the ATAC randomized trial. *Lancet.* 2002,559:2131-2139
[58] Goss PE, Ingle JN, Martino S et al. A randomized trial of letrozole in postmenopausal women after five years of tamoxifen therapy for early-stage breast cancer. *N Engl J Med.* 2003, 349:1793-1802.
[59] Goss PE, Ingle JN, Martino S et al. Updated analysis of the NCIC CTG MA.17 randomized placebo (P) controlled trial of letrozole(L) after five years of tamoxifen in postmenopausal women with early stage breast cancer. *J. Clin. Oncol.*,2004,22(Suppl. 14):847.
[60] Simpson D, Curran MP, Perry CM. Letrozole: a review of its use in postmenopausal women with breast cancer. *Drugs.* 2004,64:165-172.
[61] Bulun SE, Zeitoun K, Takayama K, Noble L, Michael D, Simpson E, et al. Oestrogen production in endometriosis and use of aromatase inhibitors to treat endometriosis. *Endo Rel Cancer* 1999;6:293-301.
[62] Meinhardt U, Mullis PE. The essential role of the aromatase /p450arom. *Semin Reprod Med* 2002;20:277-84.
[63] Casper RF. Aromatase inhibitors: New experiments or new value. *Updates in Infertility treatment 2004 Seminar in Florida* 22-24 January.
[64] Hudson Stephen. Infosheet: Letrozole. *Victoria Fertility Centre.* #207-4400 Chatterton way Victoria BC. www.victoriafertility.com
[65] Mitwally MF, Casper RF. Potential of aromatase inhibitors for ovulation and superovulation induction in infertile women. *Drugs* 2006;66(17):2149-60.
[66] Mitwally MF, Casper RF. Aromatase inhibition improves ovarian response to follicle-stimulating hormone in poor responders. *Fertil Steril* 2002 April; 77(4):776-80.
[67] Badawy A, Abdel Aal I, Abulatta M. Clomiphene citrate or letrozol for ovulation induction in women with polycystic ovary syndrome: a prospective randomized trial. *Fertil Steril* 2007 Jun 18.
[68] Bayar U, Tanriverdi HA, Barut A, Ayoglu F, Ozcan O, Kaya E. Letrozole vs clomiphene citrate in patients with ovulatory infertility. *Fertil Steril* 2006 Apr; 85(4):1045-48.
[69] Mitwally MF, Casper RF. Use of an aromatase inhibitor for induction of ovulation in patients with an inadequate response to clomiphene citrate. *Fertil Steril* 2001 Feb;75(2):305-309
[70] Begum MR, Quadir E, Begum A, Begum RA, Begum M. Role of aromatase inhibitor in ovulation induction in patients with poor response to clomiphene citrate. *J Obstet Gynaecol Res* 2006 Oct;32(5):502-506.

[71] Begum MR, Ferdous J, Begum A, Quadir E. Comparison of efficacy of aromatase inhibitor and clomiphene citrate in induction of ovulation in polycystic ovarian syndrome. *Fertil Steril* Available online 4 January 2008.

[72] Badawy A, Mosbah A, Shady M. Anastrozole or letrozole for ovulation induction in clomiphene resistant women with polycystic ovary syndrome: a prospective randomized trial. *Fertil Steril* 2007 Aug 7.

[73] Gregoriou O, Viahos NF, Konidaris S, Papadis K, Botsis D, Creatsas GK. Randomized controlled trial comparing superovulation with letrozole versus recombinant follicle-stimulating hormone combined to intrauterine insemination for couples with unexplained infertility who had failed clomiphene citrate stimulation and intrauterine stimulation. *Fertile Steril* 2007 Oct 23.

[74] Quintero RB, Urban R, Lathi RB, Westphal LM, Dahan MH. A comparison of letrozole to gonadotrophins for ovulation induction, in subjects who failed to conceive with clomiphene citrate. *Fertil steril* 2007 Oct;88(4):879-85.

[75] Jee BC, Ku SY, Suh CS, Kim KC, Lee WD, Kim SH. Use of letrozole versus clomiphene citrate combined with gonadotropins in intrauterine insemination cycles: a pilot study. *Fertil Steril* 2006 Jun;85(6):1774-7.

[76] Fatemi HM, Kolibianakis E, tournaye H, Camus m, Steirtegham Van AC, Devroy P. Clomiphene citrate versus letrozole for ovarian stimulation: a pilot study. *Reprod Biomed Online* 2003 Nov;7(5):543-546.

[77] Topipat C, Choktanasiri W, Jultanmas R, Weerakiet S, Wongkularb A, Rojanasakul A. A comparison of the effects of clomiphene citrate and the aromatase inhibitor letrozole on superovulation in Asian women with normal ovulatory cycles. *Gynecol Endocrinol* 2008 Mar;24(3):145-50.

[78] Noriega-Portella L, Noriega-Hoces L, Delgado A, Rubio J, Gonzales GF. Effect of letrozole at 2.5mg or 5.0mg/day on ovarian stimulation with gonadotropins in women undergoing intrauterine insemination. *Fertil Steril* 2007 Dec 14.

[79] Bayar U, Basaran M, Kiran S, Coskun A, Gezwe S: Use of an aromatase inhibitor in patients with polycystic ovary syndrome: a prospective randomized trial. *Fertil Steril* 2006 Nov;86(5):1447-51.

[80] Holzer H, Casper R, Tulandi T. A new era in ovulation induction. *Fertil Steril* 2006 Feb;85(2):277-84.

[81] Weil SJ, Vendola K, Zhou J et al. Androgen receptor controlled ovarian hyperstimulation in the primate ovary: cellular localization, regulation, and functional correlation. *J Clin Enocrinol Metab* 1998;83:2479-2485.

[82] Weil S, Vendola K, Zhou J, Bondy CA. Androgen and follicle stimulating hormone interactions in primate ovarian follicle development. *J Clin Endocrinol Metab* 1999;84:2951-2956.

[83] Vendola K, Zhou J, Wang J, Famuyiwa OA, Bienre M, Bondy CA. Androgens promote oocyte insulin-like growth factor 1 expression and initiation of follicle development in the primate ovary. *Biol Reprod* 1999;61:353-357.

[84] Vendola KA, Zhou J, Adesanya OO, Weil SJ, Bondy CA. Androgen stimulate early stages of follicular growth in the primate ovary. *J Clin Invest* 1998;101:2622-2629.

[85] Mitwally MF, Casper RF. Aromatase inhibition decreases FSH dose needed during controlled ovarian hyperstimulation: a controlled prospective trial [abstract 127*] J Soc Gynecol Investig* 2001;8 (Suppl): 85A.

[86] Healey S, TanSL, Biljan MM. Effects of letrozole on superovulation with gonadotropins in women undergoing intrauterine insemination. *Fertil Steril* 2003;80:1325-1329.

[87] Pellicer A, Rulz A, castellvi RM,et al. Is the retrieval of high number of oocytes desirable in patients treated with gonadotropins-releasing hormone analogues (GnRHa) and gonadotropine? *Hum Reprod* 1989; 4:536-40.

[88] Hadi FH, Chantler E, Anderson E, et al. Ovulation induction and endometrial steroid receptors. *Hum Reprod* 1994; 9-2405-10.

[89] Simon C, Cano F, Valbuena D et al. Clinical evidence for a detrimental effect on uterine receptivity of high serum estradiol concentrations in high and normal respondent patients. *Hum Reprod* 1995; 10:1432-7.

[90] Garcia JE, Acosta AA, Hsiu JG, et al. Advanced endometrial maturation after ovulation-induction with human menopausal gonadotropins/human chorionic gonadotropin for in vitro fertilization. *Fertil Steril* 1984; 42:31-5.

[91] Forman R, Fries N, Testart J et al. Evidence for an adverse effect of elevated serum estradiol concentration on embryo implantation. *Fertil Steril* 1988;49:118-22.

[92] Kolb BA, naimabadi S, Paulson PJ. Ultrastructural characteristics of the luteal phase endometrium in patients undergoing controlled hyperstimulation. *Fertil Steril* 1997;67-625-30.

[93] Paulson RI, Sauer MV, Lobo RA. Embryo implantation after human in vitro fertilization: importance of endometrial receptivity. *Fertil Steril* 1990;53:870-4.

[94] Ertzeid G, Storing R. Adverse effects of gonadotropin treatment on pre and post implantation development in mice. *J Reprod Fertil.* 1992;96:649-55.

[95] Warner CM, Cao W, Exley GE, et al. Genetic regulation of egg and embryo survival. *Hum Reprod* 1998;13(Suppl 3): 178-90.

[96] Valbuena D, Martin J, De Pabio JL, et al. Increasing levels od estradiol are deleterious to embryonic implantation because they directly effect the embryo. *Fertl Steril* :76:962-8

[97] Kim HC, Kammann E, Sheldon RM, et al. Response of blood coagulation parameters to elevated endogenous 17 beta-estradiol levels induced by human menopausal gonadotropins. *Am J Obstet Gynecol* 1981;140:807-10.

[98] Lox C, Canez M, Deleon F, et al. Hyperestrogenism induces by menotropins alone or in conjunction with leuprolide acetate in in vitro fertilization cycles:The impact of hemostasis. *Fertil Steril* 1996;63:566-70.

[99] Van der Auwera I, Pijneborg R, Koninckx PR. Yhe influence of in vitro culture versus stimulated and untreated oviductal environment on mouse embryo development and implantation. *Hum Reprod* 1999;14:2570-4.

[100] Simon C, Garcia VJ, Valbuena D et al. Increasing uterine receptivity by decreasing estradiol levels during the pre-implantation period in high responders with the US6 of a follicle-stimulating hormone step-down regimen. *Fertl Steril* 1998;70:134-9.

[101] Biljan MM, Hemmings R, Brassand N. The outcome of 150 babies following the treatment with letrozole or letrozole and gonadotropins. *Fertil Steril* 2005;84:S95.

[102] Tulandi T, Martin J, Al-Fadhli R, Kabli N, Forman R, Hitkari J, et al. Fetal safety of letrozole and clomiphene citrate for ovulation induction. *J Obstet Gynaecol Can* 2007;29(8):668-71.

In: Aromatase Inhibitors: Types, Mode of Action and Indications ISBN: 978-1-60741-711-8
Editor: Jean R. Lamonte

Chapter IV

Upfront vs. Sequential Aromatase Inhibitors: An Ongoing or Settled Debate?

Tallal Younis[1,*] and Chris Skedgel[2]

[1]Medical Oncologist, Queen Elizabeth II Health Sciences Centre. Assistant Professor of Medicine, Department of Medicine at Dalhousie University, Canada

[2]Health Economist, Department of Medicine at Dalhousie University, Canada

Abstract

Adjuvant hormonal therapy with an aromatase inhibitor has become a standard treatment option for post menopaual women with breast cancer. Compared to tamoxifen alone, upfront aromatase inhibitor and sequential tamoxifen - aromatase inhibitor strategies have been shown to improve the recurrence risk of breast cancer in a number of clinical trials. Despite the higher costs associated with aromatase inhibitors, compared to tamoxfien, both approaches have also been shown to be cost effective strategies with good value for money. As such, aromatase inhibitors are currently indicated as adjuvant hormonal therapy in post menopausal women with breast cancer considering their favorable clinical and economic profile. This chapter reviews the current evidence, from a clinical and an economic perspective, with regard to the preferred adjuvant strategy of aromatase inhibitor whether upfront or sequential treatment. The evolution of clinical evidence from indirect across studies analyses / modeling to the first results of a direct randomized study comparing both approaches is discussed. Economic evaluations based on indirect comparison across clinical studies are reviewed, and a direct head to head comparison based on BIG 1-98 study is presented.

* Corresponding author: Tallal Younis, MBBCh. FRCP (UK). Queen Elizabeth II Health Sciences Centre, 454 Bethune Building. 1278 Tower Road, Halifax, Nova Scotia. B3H 2Y9, Canada. Tel: (902) 473 6054; Fax: (902) 473 6186; E mail: tallal.younis@cdha.nshealth.ca

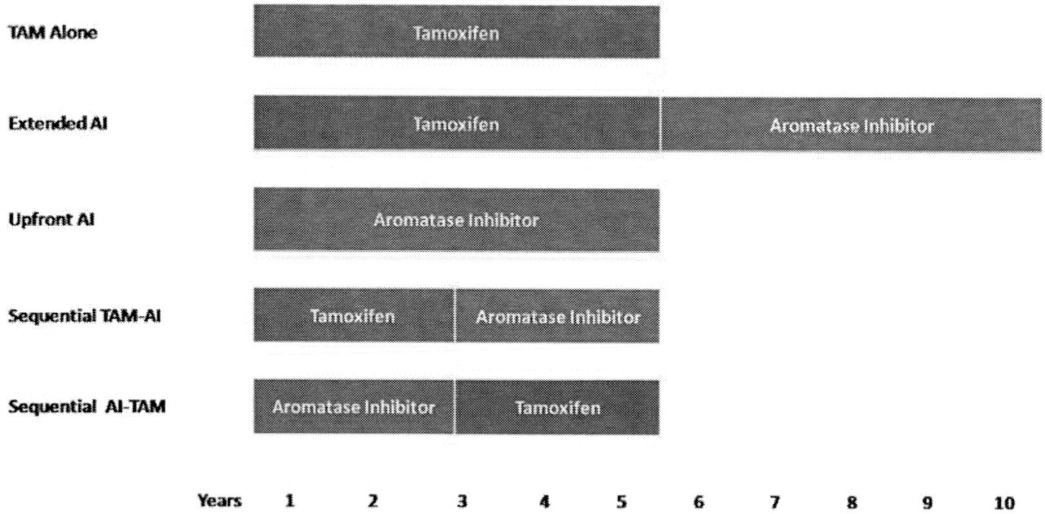

Figure 1. Hormonal Treatment Strategies.The sequence and treatment duration of tamoxifen (TAM) and/or aromatase inhibitors (AI) are shown for the hormonal treatment strategies currently employed, including Tamoxifen (TAM) alone, extended AI, upfront AI, sequential TAM-AI and sequential AI-TAM.

Introduction

Postmenopausal women with estrogen (ER) and/or progesterone (PR) receptor positive breast cancer constitute the largest group of patients with early-stage breast cancer [i]. For these women, adjuvant hormonal treatment integrating an aromatase inhibitor (AI) has emerged as a preferred strategy compared to treatment with tamoxifen (TAM) alone for five years [ii]. Currently, many postmenopausal women are undergoing adjuvant hormonal treatment with one of three available third-generation AIs {anastrazole (Arimidex®); exemestane (Aromasin®); letrozole (Femara®)} in treatment strategies that include upfront, sequential or extended approaches (Figure 1) [iii,iv,v,vi]. The upfront AI strategy involves five years of an AI following curative intent surgery, while sequential TAM-AI refers to 2-3 years of an AI following an initial 2-3 years of TAM to complete a total of five years of adjuvant hormonal therapy. A sequential AI-TAM strategy involving AI for 2 years followed by TAM for 3 years has also recently emerged as an acceptable treatment approach. In the extended AI strategy, post menopausal women with breast cancer who are disease free following 5 years of adjuvant TAM are treated with an additional five years of an AI.

A number of randomized clinical trials have demonstrated that integration of adjuvant AIs is associated with improved efficacy outcomes compared to TAM alone, but the optimal AI strategy from a clinical perspective remains undefined [iv,v,vi]. Moreover, cost implications of the different available strategies are important for decision making in health care jurisdictions given the considerably higher upfront costs associated with AIs relative to TAM [vii]. What is preferable from a clinical perspective based on efficacy outcomes may not necessarily be optimal from an economic perspective based on "value for money" (i.e. cost-effectiveness). Economic evaluations such as cost-effectiveness analyses (CEA) or cost-utility analyses (CUA) examine costs in the context of cumulative benefits, specifically life-years (LY) or quality-adjusted life-years (QALY) gained, as opposed to absolute or cumulative clinical benefits typically assessed by evaluating differences in disease-free (DFS) and overall survival (OS) outcomes [viii,ix]. Cost-effectiveness is, therefore, a relative measure of cost-efficiency that may help prioritize interventions according to their relative cost-effectiveness. Costs and gains associated with an intervention are often compared to costs and gains associated with other alternatives to determine the relative cost-effectiveness of an intervention against all comparators. Cost effectiveness analyses can also be used to judge the economic plausibility of an intervention relative to commonly accepted cost-effectiveness thresholds. Interventions associated with cost-effectiveness below acceptable thresholds are considered favourable strategies from an economic perspective, while those with cost-effectiveness above these thresholds are deemed unfavourable [x,xi]. Unfortunately, what represents an "acceptable" cost-effectiveness threshold is still a subject of considerable debate. Critical thresholds of $50,000 or $100,000 / QALY gains are commonly cited in the North America literature, and thresholds of £20,000 or £30,000 / QALY gains are commonly employed in the UK, but there is no underlying empirical or observational basis for these figures [xii,xiii].

This chapter examines the clinical and economic evidence with regard to the available AI strategies for post menopausal women with early stage hormone receptor positive breast cancer. First, we provide a brief overview of the clinical and economic studies that examined an AI-incorporating strategy compared to a TAM alone approach. Subsequently, we review a number of clinical and economic evaluations of upfront AI vs sequential TAM-AI based on indirect across-studies comparisons. Finally, we discuss clinical results and an economic evaluation based on direct head to head comparison between both strategies.

Upfront and Sequential AI Strategies

Clinical Perspective

The clinical outcomes observed in trials that examined AI strategies have been recently reviewed [iv,v,vi]. DFS and/or time-to-recurrence (TTR) outcomes were based on variable combinations of breast cancer recurrence events (i.e. distant recurrence as well as loco-regional recurrences including ipsilateral and contralateral breast cancers) with or without death, but were not defined uniformly across all studies. We herein provide a brief overview of the design and conclusions of these individual studies before comparing both strategies.

Three studies, and a recent meta-analysis, showed that upfront AI is associated with improved DFS and/or TTR outcomes compared to TAM alone [xiv,xv,xvi,xvii,xviii]. The "Arimidex Tamoxifen Alone or in Combination" (ATAC) trial examined five years of adjuvant hormonal therapy with upfront anastrazole (ANA) compared with either TAM alone or a combination of both [xiv]. There were 6241 patients randomized to ANA or TAM alone, while the third arm combining ANA and TAM was closed prematurely as it was not associated with improved outcomes relative to TAM alone. At a median follow up of 100 months, there was a 15% relative improvement in DFS (HR=0.85, 95%CI: 0.76-0.94) and a

24% relative improvement in TTR (HR=0.76, 95%CI: 0.67-0.87) with upfront ANA compared to TAM alone in the approximately 84% patients with hormone receptor positive disease. No OS benefit was observed. The "Breast International Group" (BIG) 1-98 was a four-arm trial comparing TAM alone, upfront letrozole (LET), sequential LET-TAM and sequential TAM-LET (Figure 2) [xv,xvi]. There were 1828 patients randomized to the 2- arm option (TAM alone vs upfront LET), and 6182 patients randomized to the 4-arm option (TAM alone, upfront LET, TAM-LET and LET-TAM). The monotherapy arm analysis with a median follow up of 51 months demonstrated improvements in DFS and distant disease free survival (DDFS) with upfront LET compared to TAM alone [xv]. Subsequently, approximately one third of the patients switched to an AI following unblinding of the TAM alone arm. An updated analysis with a median follow up of 76 months that included all patients in the monotherapy arms was recently presented [xvi]. Both intention to treat and censored analyses were presented to account for the impact of unplanned cross over from TAM to an AI after unblinding of the TAM alone arm. In this analysis upfront LET was associated with significant improvements in DFS in both the intention to treat (HR=0.84, 95%CI: 0.74-0.95) and censored (HR=0.88, 95%CI: 0.78-0.99) analyses. There was also a trend for improved OS (HR=0.87, 95%CI: 0.75-1.02). The "Tamoxifen Exemestane Adjuvant Multinational" (TEAM) study also showed improvements in DFS with upfront AI relative to TAM alone [xvii]. Although TEAM was initially designed to examine upfront exemestane (EXE) vs TAM alone, it was revised in 2004 to compare upfront EXE to sequential TAM-EXE after the "Intergroup Exemestane Study" (see below) showed improved outcomes with the TAM-EXE strategy relative to TAM alone [xix]. In the first planned analysis of TEAM, at 2.75 years, all women were censored at 2.75 years before the planned switch from TAM to EXE in the TAM-EXE arm. This analysis therefore reported outcomes for upfront EXE relative to TAM alone with a relatively short median follow up. In an intention to treat analysis, upfront EXE was associated with a non-statistically significant improvement in DFS relative to TAM (HR=0.89, 95%CI: 0.77-1.03). A recently reported meta-analysis conducted by the Early Breast Cancer Trialists' Collaborative Group (EBCTCG) that included 9,856 patients with 50,000 woman-years of follow-up, from ATAC and BIG 1-98, examined upfront AI monotherapy compared to TAM alone [xviii]. The upfront AI strategy was associated with a 23% relative reduction (HR=0.77, SE 0.05) in the risk of breast cancer recurrences (i.e. local, contralateral and distant), corresponding to an absolute reduction of 2.9% at 5-years and 3.9% at 8 years. Non significant reductions in breast cancer mortality of 1.1% at 5-years and 0.5% at 8 years were also observed. There appeared to be a greater proportional decrease in isolated local or contralateral recurrence rates compared to distant recurrence rates, but no apparent heterogeneity was noted in other subgroups with respect to age, nodal status, tumor grade or PR receptor status. These clinical randomized studies along with the EBCTCG meta-analysis collectively show that the upfront AI strategy is associated with significant improvements in breast cancer recurrences compared to TAM alone.

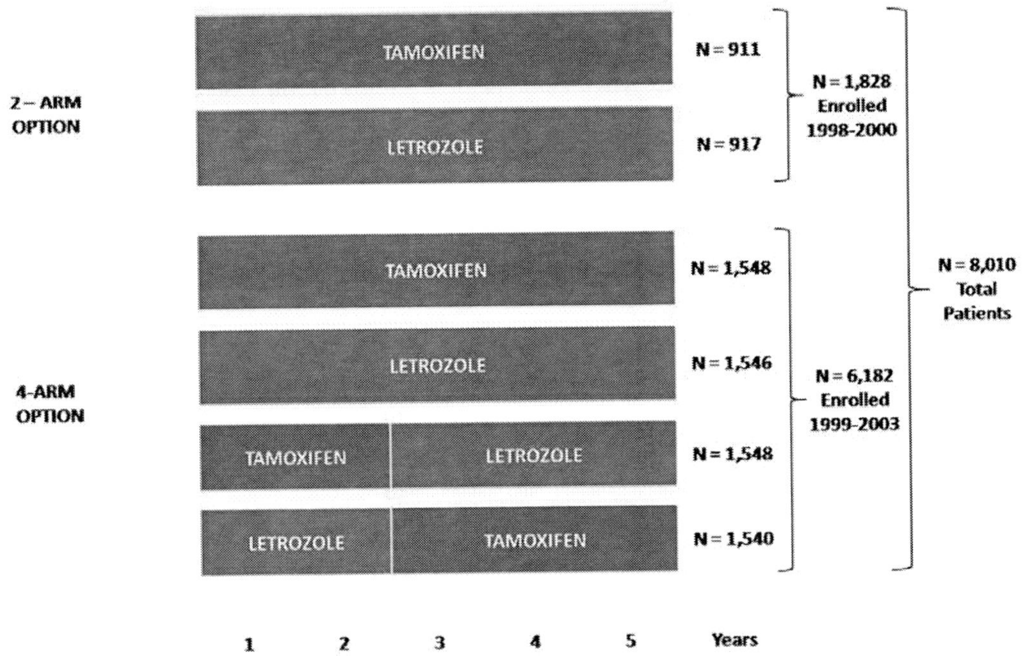

Figure 2. BIG 1-98 Study Schema.Treatment strategies employed, and the number of patients enrolled, in the 2- arm and 4-arm options of BIG 1-98 are shown.

Three studies, and two meta-analyses, also demonstrated that sequential TAM-AI is associated with improved DFS and/or TTR compared to TAM alone [xviii,xix,xx,xxi,xxii,xxiii,xxiv]. The "Intergroup Exemestane Study" (IES) included 4724 women treated with adjuvant TAM for 2–3 years who were randomized to either continue TAM or switch to EXE for 2–3 years to complete 5 years of hormonal therapy [xix]. A smaller study, the "Italian Tamoxifen Anastrozole" (ITA) trial, randomized 448 women with node-positive breast cancer who remained disease free after 2 to 3 years of TAM treatment to either continue TAM or cross over to ANA [xx]. Finally the "Austrian Breast and Colorectal Cancer Study Group" (ABCSG) Trial 8 / "Arimidex-Nolvadex" (ARNO) 95 trial randomized 3224 women after definitive surgery to either sequential ANA after 2 years of TAM or to TAM alone for a total of 5 years [xxi,xxii,xxiii]. Each of these studies observed significant improvements in DFS with either a cross-over (IES and ITA) or a planned switch (ABCSG/ARNO) strategy relative to TAM alone. The magnitude of benefit, however, varied across studies with hazard ratios (HR) for DFS ranging between 0.35 (95%CI: 0.18-0.68) in the ITA study and 0.75 (95%CI: 0.65-0.87) in the larger IES trial [xix,xx]. A meta-analysis of the latter three studies (i.e. ITA, ABCSG 8 and ARNO 95) reported statistically significant improvements in DFS (HR=0.59, 95%CI: 0.48-0.74) and OS (HR=0.71, 95%CI: 0.52-0.98) favoring sequential TAM-AI relative to TAM alone [xxiv]. A subsequent and larger meta-analysis conducted by the EBCTCG, including 9,015 patients with 33,000 woman-years of follow-up, examined AI after 2-3 years of TAM as a switch strategy compared to TAM alone [xviii]. Sequential AI was associated with a 29% relative reduction (HR=0.71, SE 0.06) in the risk of breast cancer recurrence corresponding to an absolute reduction of 3.1% at 3-years and 3.5% at 6-years from treatment divergence. Significant reductions in breast cancer mortality of 0.7% at 3-years and 1.6% at 5-years were also observed. There again appeared to be greater proportional decreases in isolated local or contralateral disease recurrence rates compared to distant recurrence rates, but no apparent heterogeneity was noted in other subgroups. Although BIG 1-98 trial (see above) included four treatment strategies (TAM alone, upfront LET, sequential TAM-LET, sequential LET-TAM), a direct comparison between the sequential arms and TAM alone has not been reported [xv,xvi]. Instead, a direct comparison of the sequential arms compared to upfront AI was presented considering that upfront LET was already shown to be associated with improved DFS relative to TAM alone.

Clinical trials also support extended therapy with AI following five years of TAM alone. The MA-17 trial, led by the National Cancer Institute of Canada (NCIC), randomized 5187 women who were disease free following 5 years of adjuvant TAM to either extended LET for five years or placebo [[xxv]]. Extended LET was associated with improved DFS in all patients (HR=0.58, 95%CI: 0.45-0.75) and OS in the 46% of patients with node positive disease (HR=0.61, 95%CI: 0.38-0.98).

Economic Perspective

Cost-effectiveness analyses of adjuvant AI strategies in post menopausal women with breast cancer has been previously reviewed [[xxvi],[xxvii]]. In these analyses, the incremental costs associated with AI strategies compared to TAM alone were associated with improved outcomes in terms of fewer recurrences and higher QALY gains. Most importantly, AI strategies were associated with favourable cost-effectiveness relative to TAM alone in different jurisdictions/countries. A review of thirteen economic evaluations of adjuvant treatments incorporating AIs also identified a number of key factors with strong impact on cost-effectiveness results or those considered of key importance according to modelling guidelines [xxvii]. These included: (1) relative risk reduction of recurrences with AI treatment, (2) absolute incidence of recurrence in the control arm, (3) patient age, (4) cost per day of the intervention including hormonal drugs and routine follow-up costs, and (5) the total cost of adverse events. The first three factors largely determine the magnitude of clinical benefit achieved with AI therapy (i.e. effectiveness) while the latter two influence the cost component of the evaluation.

Efficacy models, whether clinical or economic, require extrapolation of treatment effects and toxicities beyond the relatively short follow up period in clinical trials [viii,ix]. AI cost-effectiveness evaluations have involved modeling for long term efficacy outcomes (e.g. 10-year to lifetime) from the short term efficacy results reported by the available clinical trials (e.g. 2 to 5 years) [xxvi,xxvii]. In these analyses, the treatment effect (i.e. magnitude of benefit) was dependent on the baseline risk of cancer recurrence without AI treatment, the observed relative benefit with AI treatment and the carry over benefit for AI beyond treatment duration. The uncertainties with regard to each of these variables add significant methodological challenges to the analyses and/or constitute limitations to the robustness of the results, which have been acknowledged in these reports. Firstly, AI treatments were prescribed to heterogeneous cohorts of patients with varying baseline recurrence risks ranging from patients with low-risk node-negative disease to patients with high-risk node-positive disease. Secondly, the point estimates for treatment effect (e.g. HR for recurrence) are volatile over time within studies and/or at relevant time points across studies. Thirdly, although AI trials have observed a carry-over benefit for the AI strategy beyond treatment duration, similar to what has been observed with TAM, the duration and magnitude of the treatment effect over varying time horizons remains undefined [xxviii]. Despite these limitations, conclusions with regard to the overall cost-effectiveness of adjuvant AI therapy for postmenopausal women with hormone-receptor positive breast cancer have been robust, in sensitivity analyses, across plausible ranges of uncertainties for these parameters in all economic evaluations reported to date [xxvi,xxvii].

Indirect Across-Studies Comparisons

The individual clinical trials and economic analyses discussed above have observed that AI strategies are associated with improved clinical outcomes and favourable cost-effectiveness relative to TAM alone for post menopausal women with breast cancer. These studies, however, have not provided a clear direction as to whether an upfront or sequential AI strategy is optimal. A number of evaluations have attempted to clarify this question based on indirect across-studies comparisons from a clinical perspective with regard to improved clinical outcomes or an economic perspective with reference to acceptable cost-utility thresholds [xxix,xxx,xxxi,xxxii,xxxiii,xxxiv,xxxv]. The AI strategies examined (upfront vs sequential as well as the extended strategy in certain studies) were based on the treatment schedule employed in the relevant clinical trials (Figure 1). As upfront, sequential and extended AIs have all been examined compared to five years of TAM alone, these evaluations were primarily based on indirect comparisons using TAM alone as a common comparator. All models accounted for early relapses that could occur while on TAM and before switching to an AI. Models, however, are simplified representations of reality. They cannot incorporate all possible sequences of events, and must rely on key assumptions to reduce an infinitely complicated process to a manageable analysis [viii,ix]. The clinical and economic models herein examined are no exception. In all models sensitivity analyses have been required, to test the robustness of results within reasonable ranges of uncertainty around key parameters and/or assumptions. A number of these models have also provided one- or two-way sensitivity analyses to determine the optimal AI strategy for future analyses based on updated and/or ongoing AI clinical trials [xxix,xxxii,xxxiii,xxxiv].

Clinical Perspective

Punglia et al and Cuzick et al have examined 10-year efficacy outcomes, including DFS and/or TTR, associated with AI strategies based on short term follow up results from relevant AI clinical trials [xxix,xxx].

Punglia et al developed a Markov model to simulate the clinical history for postmenopausal women with hormone receptor-positive breast cancer treated with: (1) TAM alone, (2) Upfront AI, (3) Sequential TAM-AI with a cross over at 2.5-years, and (4) Extended TAM-AI with cross over at 5-years [xxix]. Three main health states were included in the model: (1) well with no evidence of cancer recurrence, (2) recurrent local or regional disease including a new primary breast cancer (either ipsilateral or contralateral), or (3) metastatic disease. Efficacy outcomes were primarily based on HRs for DFS from relevant AI studies: 1) 0.82 for upfront AI based on the ATAC study, 2) 0.67 for TAM-AI based on the IES trial, and 3) 0.57 for extended AI based on the MA 17 study [xiv,xix,xxv]. A carry over benefit for AI was assumed for the 10 year horizon duration examined with magnitude based on the HRs for DFS. Separate models were developed for patients with node negative and node positive breast cancer to assess the impact of varying baseline risks. DFS and distant disease-free survival (DDFS) as well as average time spent without disease and average time spent without distant disease were all calculated at a 10-year time horizon. Overall survival was not modelled.

In this analysis [xxix], the sequential TAM-AI was associated with a modest improvement in 10-year DFS and DDFS relative to upfront AI in patients with node-negative or node-positive disease. The model predicted a 6% relative risk improvement in DFS events at 10-years for sequential TAM-AI compared to upfront AI, which translated into a 1.1% and 2.1% absolute risk reduction in DFS for node negative (82.6% vs 83.7%) and node positive (65.5% vs 67.6%) disease, respectively. Conversely, quantitative estimates of DFS (i.e. the cumulative 10-year area under the curve of patient-months free of recurrence) slightly favoured the upfront AI strategy by 0.1 patient months (i.e. +3 days) as a result of the lower likelihood of early recurrence with the upfront AI strategy. Extended TAM-AI strategy did not improve 10-year DFS estimates further. These conclusions, with regard to the rankings of AI strategies, were robust to key parameters in the model including the degree of carry over benefit assumed. The investigators, however, cautioned that model outcomes based on short-term follow-up results, such as those in the AI studies, may not accurately reflect long term outcomes as the optimal AI strategy choice in the model was not apparent until after 5 years of initial therapy. As well, a two-way sensitivity analysis incorporating a plausible range of efficacy HRs for AI strategies was provided to inform future analyses based on updated results from ongoing or new AI studies. If point estimates of HRs from the recently reported EBCTCG meta-analyses for upfront (i.e. HR=0.77, SE 0.05) and sequential AI (i.e. HR=0.71, SE 0.06) were examined in the two-way sensitivity graphs provided by Punglia et al, the upfront AI strategy would appear to be associated with improved absolute DFS estimates relative to a sequential AI strategy. These results, however, are not robust across the 95% CI range of uncertainty around the point estimates of HRs despite the inclusion of more than 19,000 patients in this meta-analyses resulting in relatively narrow 95% CIs [xviii].

Cuzick et al also attempted to predict 10-year breast cancer recurrence risk, as well as TTR, associated with different AI strategies [xxx]. The efficacy end points examined were "recurrence rates" at a 10-year horizon and "time lost to recurrences" within a 10-year time frame. Three strategies were examined: (1) Upfront AI, (2) Sequential TAM-AI with a cross over at 2-years, and (3) Extended TAM-AI with cross over at 5-years. The efficacy outcomes were primarily based on HRs for TTR, as opposed to HRs for DFS in Punglia et al, from the relevant AI studies: 1) 0.73 for upfront AI beginning at onset of treatment, 2) 0.65 for TAM-AI beginning 2-years after initial treatment with TAM, and 3) 0.60 for extended AI beginning

5-years after initial treatment with TAM. A carry over benefit for AI was assumed for another five years beyond AI treatment in the primary analysis but models with a 2-year efficacy carryover were also evaluated. Efficacy outcomes were, overall, in favour of an upfront AI strategy compared to a sequential TAM-AI strategy although incremental gains were small and model-dependent. In the primary analysis with a 5-year carry over benefit assumption, upfront AI was associated with a slightly worse absolute recurrence risk estimate (17.3% vs 17.1%) but a 0.7 month (i.e. +21 day) incremental gain in time lost to recurrence estimate compared to sequential TAM-AI (11.5 vs 10.8 months) at the 10-year horizon examined. The extended AI strategy was associated with inferior outcomes relative to upfront and sequential approaches.

At first glance, these two analyses may appear to have reached different conclusions with regard to the preferred AI strategy from a clinical perspective [xxix,xxx]. However, a number of conclusions could be drawn from both analyses. Firstly, the difference in absolute recurrence rates between upfront and sequential AI strategies predicted by both analyses at a 10-year horizon is very small if present at all (i.e. between 0 to 2%). Secondly, the upfront AI strategy appeared to be associated with small incremental gains in "life years" or "recurrence-free years" compared to the sequential TAM-AI in both analyses. The latter outcome (i.e. life year or recurrence-free year) seems more meaningful compared to the former one (i.e. absolute recurrence risk or absolute DFS) as they give more weight to early recurrences within the time frame examined. Incremental gains or losses in LYs and QALYs are a function of both absolute number of events (e.g. cancer recurrence) and time, where earlier recurrences translate into more significant losses in LYs and QALYs compared to later recurrences. Thirdly, the magnitude of benefit derived depends on the predicted baseline risk of cancer recurrence. Patients with higher baseline recurrence risks (i.e. node positive) derive more incremental benefits compared to patients with lower recurrence risks (i.e. node negative). Fourthly, the predicted small incremental benefits with regard to efficacy highlight the importance of other outcomes, such as adverse events of hormonal treatment and/or costs, for decision making [v,vii]. Although neither was examined in these analyses, both are important domains without which treatment decisions cannot be accurately made. As an upfront AI strategy is associated with higher upfront drug costs compared to a sequential TAM-AI strategy, it remains unclear whether these higher incremental drug costs are justified considering the modest, if any, incremental gains in LY or QALY associated with upfront relative to sequential AI. In economic terms, "is an upfront AI strategy cost-effective relative to a sequential TAM-AI strategy?"

Economic Perspective

A number of economic evaluations have been conducted to examine the cost per incremental benefit (i.e. cost-effectiveness) associated with AI strategies as opposed to absolute or cumulative benefit alone (i.e. efficacy outcomes) [xxxi,xxxii,xxxiii,xxxiv,xxxv]. These economic analyses have been based on indirect across-studies comparisons, and are described below.

Lønning constructed an economic model to compare cost utilities, in terms of cost per LY and QALY gained, associated with different AI strategies based on hypothetical assumptions for efficacy rather than clinical trial data [xxxi]. Health states included in the model were hormonal treatment, relapse-free (with or without complication), relapse (including contralateral breast) and death. The model assumed that all AI strategies are associated with similar absolute benefits compared to TAM alone, corresponding to a 2% absolute improvement in recurrence and a 1.5% absolute improvement in overall survival. The constant absolute reduction in recurrence rate, assumed for all strategies, conforms to the larger reduction in HRs observed in clinical trials for extended vs sequential vs upfront AI strategies relative to TAM alone. Separate cohorts, with node positive and node negative disease, were examined. No differences in costs between AI drugs or in adverse events among AI strategies were assumed. Costs and benefits were discounted at 3%, and sensitivity analyses were performed for key parameters.

Under the assumption of an equivalent 2% improvement in recurrence risk for all AI strategies, sequential TAM-AI was associated with the lowest cost/QALY estimate while extended AI was associated with the highest cost. Sensitivity analyses for the impact of age, nodal status, fracture risk, biphosphonate cost and relapse costs influenced the cost-utility estimates but did not change the rankings of AI strategies. Most importantly, Lønning observed that an upfront AI strategy would be associated with a moderate cost/QALY gains compared to sequential TAM-AI if it resulted in an additional 1% improvement in absolute recurrence risk (i.e. 3% reduction in recurrence risk with upfront AI vs 2% reduction with TAM-AI). The cost-utility of upfront AI relative to sequential TAM-AI varied according to age and nodal status (node positive vs node negative) and was between US$ 22,500-30,000/QALY for patients 55 years of age, US$ 29,000-40,000/QALY for patients 65 years of age and US$ 43,500-62,000/QALY for patients 75 years of age.

Skedgel et al conducted an economic evaluation to explicitly compare the cost utilities of upfront and sequential AI strategies based on early reports from the ATAC and IES trials [xxxii]. A state-transition model was constructed to examine costs and QALY gains associated with upfront ANA or sequential TAM–EXE relative to TAM alone at 10- and 20-year horizons in a cohort of hypothetical patients, 39% of whom had node positive disease as observed in relevant studies. The model incorporated five health states: (1) well on treatment, (2) well off treatment, (3) local relapse (combining locoregional and contralateral recurrences), (4) distant relapse and (5) death (with or without relapse). Adverse events modeled were derived from the relevant clinical studies and included vaginal bleeding, endometrial carcinoma, venous thromboembolism and bone fractures. The efficacy outcomes for upfront and sequential AI strategies were based on HRs of disease recurrence for women with ER positive tumours in the ATAC and IES studies, respectively. Although HRs based on TTR rather than DFS would have been preferred, they were not reported consistently across all studies. As such, the model relied on HRs for DFS (0.83 vs 0.70; ATAC vs IES, respectively) in the primary analysis and on derivation of HRs for TTR (0.74 vs 0.58) in a secondary analysis. A five-year carry over benefit for AIs beyond the initial five-year hormonal therapy was assumed. Costs and outcomes were discounted by 3%.

AI strategies were associated with higher upfront drug costs compared to TAM alone. Based on monthly costs of Cdn$ 10.5 for TAM and Cdn$ 161 for AI, a five-year treatment course would cost Cdn$ 630 for TAM alone, Cdn$ 5149 for sequential TAM-AI and Cdn$ 9668 for upfront AI. AI strategies were also associated with lower recurrence risk estimates and cumulative QALY gains relative to TAM alone. In the primary analysis, the model projected a 2.4% and 3.3% improvement in DFS at 5-years for upfront ANA and sequential TAM-EXE strategies, respectively relative to TAM alone corresponding to a 0.9% differential in favour of sequential TAM-EXE. As observed in the models by Punglia et al and Cuzick et al [xxix,xxx], however, the upfront ANA strategy was associated with modest QALY gains of +6 to +13 days at 10-years compared to sequential TAM-EXE in the primary and secondary analyses, respectively. Nevertheless, upfront ANA was associated with unfavourable incremental costs per QALY gains relative to sequential TAM-EXE in the context of a Cdn$ 50,000/QALY gained threshold.

In a later report, Skedgel et al adapted the original Canadian analysis to a Belgian perspective, in order to assess the generalizability of the model findings in another health care jurisdiction, and added an extended AI strategy to the analysis [xxxiii]. Efficacy outcomes were based on the updated HRs for recurrence (i.e. DFS in the primary analysis and TTR in the secondary analysis) for AI strategies from the ATAC, IES and MA 17 studies applied to the baseline recurrence risk in the TAM alone arm. A five-year carry over benefit for AIs beyond the initial five-year hormonal therapy was also assumed. Drug costs were based on monthly costs of € 21 for TAM and € 134 for AIs. Costs and outcomes were discounted by 3%. At a 5-year horizon, TAM-EXE was associated with a small + 0.4% absolute DFS benefit compared to upfront ANA corresponding to a 2.9% and 2.5% improvement in DFS for the sequential and upfront AI strategies relative to TAM alone, respectively. At a 20-year horizon, upfront ANA was associated with QALY gains of -6 to +30 days relative to sequential TAM-EXE in the primary and secondary analyses, respectively. Overall, sequential TAM-EXE appeared to be the economically preferred strategy at a 20-year horizon relative to € 30,000/ QALY gained threshold. Compared to sequential TAM-EXE, upfront ANA was dominated in the primary analysis and was associated with a marginal CU of € 33,786/ QALY gained in the secondary analysis. The extended strategy was also dominated by sequential TAM-EXE.

The consistent cost-effectiveness results in the Canadian and Belgian evaluations by Skedgel et al suggest that the relative cost-effectiveness of AI strategies is not primarily dependent on jurisdiction-based costs [xxxii,xxxiii]. Rather, key drivers of cost-effectiveness in both analyses were the effectiveness outcomes derived from the HR of recurrences for the different AI strategies relative to TAM alone. As such, two-way sensitivity analyses incorporating varying HRs were provided in both evaluations to determine the optimal cost-effective strategy at a 20-year horizon relative to commonly accepted cost-utility thresholds (i.e. Cdn$ 50,000/QALY in Canada and € 30,000/ QALY in Belgium) according to varying HR combinations. HRs examined for upfront and sequential AI strategy included all HRs between the point estimates of DFS and TTR hazards for upfront (0.83 vs 0.74) and sequential (0.70 vs 0.58) strategies assuming these would represent upper and lower limits of likely HRs. In the two analyses, the majority of HR combinations examined favoured the sequential strategy [xxxii,xxxiii]. Incorporating the HRs from the recently reported EBCTCG meta-analyses for upfront (HR=0.77, SE 0.05) and sequential AI strategies (HR=0.71, SE 0.06) in the two-way sensitivity analyses provided by Skedgel et al, however, show mixed results [xviii]. HR combination of point estimates (i.e. 0.77 and 0.71) almost falls on the cost-effectiveness threshold line denoting marginal cost-effectiveness for upfront AI relative to sequential TAM-AI. As well, HR combinations based on all possible estimates across the 95% CIs as opposed to point estimates show TAM-AI as the optimal cost-effective strategy in no more than half of the determinations.

Younis et al utilized a Markov model framework to predict an optimal AI strategy according to baseline risk of cancer recurrence [xxxiv]. Efficacy outcomes were primarily based on the relative benefits of AI strategies in clinical trials (i.e. HRs for recurrence) and the baseline risks of cancer recurrence without adjuvant systemic therapy. Given a constant relative benefit of a therapeutic strategy, the derived cumulative benefits and resultant cost-effectiveness are greater for disease with the highest baseline risk of recurrence. Four hypothetical cohorts with variable 10-year baseline risks of cancer recurrence were, therefore, examined to reflect the spectrum of breast cancer risk encountered in clinical practice; (i) low risk: node-negative disease with a 25% projected disease recurrence risk at 10 years; (ii) average risk: reflecting a mixed cohort of node-negative (60%) and node-positive (40%) disease with an aggregate estimated risk for disease recurrence of 35%; (iii) high risk: reflecting an estimated 50% risk of recurrence for node-positive disease and (iv) very high risk: representing node-positive patients with a high nodal burden and a 75% risk of disease recurrence. All recurrences were included including loco-regional, contralateral and distant recurrences. The primary analysis was based on a 10-year horizon, did not incorporate adverse events and judged cost-effectiveness at a 50,000/QALY gained threshold. The impact of costs, adverse events, age, analysis horizon, cost-effectiveness threshold and carry over benefit were also examined.

The evaluation provided a two-way sensitivity analyses to help predict cost-effective strategies according to baseline risk of cancer recurrence without hormonal treatment and the HRs of recurrence with an AI strategy. As clinical models often provide adjuvant treatment recommendations, including those involving hormonal therapies, based on 10-year recurrence risk without treatment, a similar approach was undertaken in this analysis [[xxxvi]]. Based on an HR estimate of 0.73 for upfront AI (average of 0.74 from ATAC and 0.72 from BIG 1-98) and 0.60 for sequential TAM-AI (derived from ARNO/ABCSG), an upfront AI strategy appeared to be cost-effective in patients with the highest baseline recurrence risk (i.e. very high risk cohort) while sequential TAM-AI was the preferred cost-effective strategy in patients with lower risk (i.e. low to average risk cohorts). Cost-effective estimates, and optimal AI strategies for patients with recurrence risk in the mid range (i.e. high risk cohort with ~ 50% recurrence risk) were dependent on other parameters including time horizon, cost-effectiveness threshold and carry over benefit assumptions. Incorporating the HRs for upfront (HR=0.77, SE 0.05) and sequential AI (HR=0.71, SE 0.06) from the EBCTCG meta-analyses [xviii] in the two-way sensitivity analyses graphs provided by Younis et al also confirms the key role for baseline recurrence risk in determining the optimal cost-effective strategy, as upfront AI was cost-effective in patients with high to very high risk while sequential TAM-AI was the preferred cost-effective strategy in patients with low to average risk.

Younis et al presented an updated economic evaluation, with a longer 25-year horizon, to examine the cost-effectiveness of adjuvant hormonal therapy in post menopausal women with breast cancer undergoing 5-years of treatment with TAM alone, an upfront AI, or sequential TAM-AI [xxxv]. Efficacy outcomes were derived from the EBCTCG meta-analyses for TAM and AIs [xviii,xxviii], with 10-year duration of benefit (i.e. 5 year treatment plus 5 year carry over benefit) assumed for all strategies. Different cohorts with varying 10-year baseline recurrence risks without adjuvant hormonal treatments were examined. No adverse events secondary to hormonal treatments were included in the primary analysis, and cost-effectiveness was based on a Cdn$ 50,000/QALY gained threshold. Costs and benefits were discounted at 3%. The analysis confirmed that all adjuvant hormonal treatments (i.e. TAM and/or AI) were cost-effective strategies relative to no adjuvant therapy. As well, cost-effective estimates were more favorable in patients with higher compared to lower disease recurrence risk. As in previous reports, the upfront AI strategy was associated with higher costs and more QALY gains compared to TAM-AI in all cohorts [xxxii,xxxiii,xxxiv]. The optimal cost-effective strategy at the 25-year horizon examined was again dependent on the baseline recurrence risks without hormonal treatment. The cost-effectiveness of upfront AI relative to sequential TAM-AI was favorable in patients with average to very-high risks but unfavorable in patients with low risks.

These economic evaluations, and indirect across-studies comparisons, reveal a number of key observations [xxxi,xxxii,xxxiii,xxxiv,xxxv]. Firstly, upfront AI is associated with higher incremental costs relative to sequential TAM-AI due to the higher upfront drug costs associated with AIs relative to TAM. These incremental costs should, however, be examined in the context of all incremental benefits achieved to assess relative “value for money”. Secondly, upfront AI strategies were associated with small, if any, incremental QALY gains compared to sequential TAM-AI strategies in most evaluations. Although clinical evaluations focus on absolute gains at relevant time points (e.g. 10-year DFS), economic evaluations are driven by cumulative gains over time (i.e. QALY gains). Thirdly, and perhaps most importantly, the key driver for cost-effectiveness of upfront relative to sequential AI strategies was the magnitude of benefits achieved with each relative to TAM alone. Given how small the potential QALY gains associated with one AI strategy relative to another may be, small changes in magnitude of benefits with either will affect conclusions in regard to optimal AI strategies. In the model by Lønning et al for example, a further 1% improvement in recurrence rate with upfront AI compared to sequential TAM-AI strategies was associated with moderate cost-effectiveness for upfront AI relative to sequential TAM-AI [xxxi]. Fourthly, the magnitude of benefit associated with AI strategies is volatile at relevant time points across studies and over time within studies. The EBCTCG meta-analyses of upfront and sequential AI strategies including approximately 19,000 patients, however, provide more definite point estimates for efficacy (i.e. relative and absolute recurrence risks) with narrower ranges of uncertainty (i.e. 95%CI) due to the large sample size included [xviii]. Fifthly, the predicted cumulative QALY gains associated with upfront relative to sequential strategies were greater in patients with higher compared to lower risks of cancer recurrence and in younger compared to older patients. Given constant relative benefits for AI strategies across patient subgroups [xviii], the cost-effectiveness of upfront relative to sequential AI strategies will vary in different clinical scenarios according to treatment benefits expected. In the evaluations by Younis et al, upfront AI was the optimal CE strategy in patients with the higher risks of recurrence while sequential TAM-AI was the preferred CE strategy in patients with the lower risks [xxxiv,xxxv].

Direct Head to Head Comparisons

The conclusions derived from modeling, whether clinical or economic, are perhaps overshadowed by methodological limitations and/or challenges that are inherent to all indirect comparisons across different studies [xxvii]. Clinical trials and economic evaluations based on direct head to head comparisons are, therefore, important in determining ultimate preferred therapeutic strategies from both clinical and economic perspectives.

Clinical Perspective

BIG 1-98 randomized 1828 women to a 2-arm option (TAM alone vs upfront LET) between 1998 and 2000, and 6182 women to a 4-arm option (TAM alone, upfront LET, sequential TAM-LET and sequential LET-TAM) between 1999 and 2003 (Figure 2). The updated results of the monotherapy arms confirmed that upfront LET is associated with an improved DFS compared to TAM alone [xvi]. Efficacy outcomes based on direct head to head comparisons among the sequential strategies and upfront LET in BIG 1-98 were also reported (Table 1), although no direct comparison between the sequential arms and TAM alone was presented [xvi]. There was a trend for improved DFS with a 1.8% absolute reduction in 5-year breast cancer events (7.3% vs 9.1%) in the upfront LET compared to the sequential TAM-LET arm. A subgroup analysis according to nodal status observed greater incremental benefits for upfront LET vs sequential TAM-LET in the node positive compared to the node negative patient cohort with absolute DFS benefits of 2.3% and 1.4%, respectively. Sequential LET-TAM and upfront LET appeared to be associated with comparable breast cancer event rates (7.3% vs 7.3%) at 5-years. Adverse events and quality of life were not reported.

Another prospective phase III trial (i.e. TEAM study) randomized 9775 women to upfront EXE or sequential TAM-EXE [xvii]. The first planned analysis at a median follow-up time of 2.75 years reported outcomes associated with the upfront EXE arm compared with TAM alone before switching to EXE in the TAM-EXE arm. The final planned analysis at a median 5-year follow up time will report the direct head to head comparison of upfront EXE compared with sequential TAM-EXE. Perhaps the larger sample size in TEAM will provide more power for the study to confirm or refute model-predicted and/or trial-observed small incremental gains among AI strategies. In the meantime, efficacy outcomes reported to date in BIG 1-98 suggest that upfront AI strategies may be associated with small incremental gains relative to TAM-AI strategies [xvi]. These non-statistically significant results, however, do not establish a clinically superior strategy. Moreover, treatment decisions will also depend on direct comparisons of adverse events and quality of life across these strategies, which remain to be reported. Longer follow up with more mature date from BIG 1-98 may provide conclusive evidence in support of one strategy over another. Conversely, this study may eventually prove to have been underpowered to confirm predicted small gains, if any, among AI strategies.

Table 1. BIG 1-98 Sequential vs Upfront AI Strategies

Hormonal Strategies	Upfront AI	TAM-AI	AI-TAM
Patients Numbers	1546	1548	1540
Disease Free Survival @ 5-Years	87.9%	86.2%	87.6%
Hazard Ratio (99% CI)	-	1.05 (0.84-1.32)	0.96 (0.76-1.21)
Breast Cancer Events @ 5-Years			
All Patients*	7.3%	9.1%	7.3%
Node-negative disease	3.5%	4.9%	3.9%
Node-positive disease	12.4%	14.7%	12.5%

Median F/U = 71 months. TAM: Tamoxifen. AI: Aromatase Inhibitor. *42% of patients had node-positive disease.

Economic Perspective

Skedgel et al conducted an economic evaluation of AI strategies based on updated results from BIG 1-98 to determine an optimal cost-effective approach in the absence of unequivocal superior clinical outcomes [xxxvii]. The analysis used a state-transition model to calculate cumulative costs and QALYs over a 25-year horizon for hypothetical cohorts of post-menopausal women with hormone receptor positive breast cancer undergoing adjuvant hormonal treatment. As the sequential arms (i.e. TAM-LET and LET-TAM) were not directly compared to TAM alone, it was not possible to directly compare all strategies. As such, the analysis conducted direct within-arm comparisons and an indirect between-arm comparison. Efficacy outcomes were derived from BIG 1-98 based on HR point estimates (Table 1) [xvi]. No adverse events were included as these have not yet been reported. Costs and outcomes were discounted at 3%.

The evaluation confirmed the favorable cost-effectiveness for upfront AI relative to TAM alone based on the monotherapy arm results. The direct within-arm comparisons demonstrated that sequential LET-TAM was associated with QALY gains and cost savings relative to upfront LET and sequential TAM-LET (i.e. LET-TAM dominated upfront LET and TAM-LET). In the indirect four-arm comparison, sequential LET-TAM also dominated upfront LET and sequential TAM-LET, and was associated with favorable cost-effectiveness relative to TAM alone. Collectively, these direct and indirect analyses suggest that sequential AI-TAM may be a dominant strategy over upfront AI and sequential TAM-AI, if longer follow up from BIG 1-98 confirms these early clinical data. It is, however, important to remember that these preliminary results were based on a deterministic approach (i.e. point estimates of efficacy HRs) rather than a probabilistic one incorporating all uncertainties around the point estimates of efficacy HRs (i.e. 95%CI). Moreover, a complete analysis should incorporate adverse events and quality of life data, which are not currently available. Further clinical data from BIG 1-98 and formal sensitivity analyses in the economic evaluation incorporating all outcomes are, therefore, required before definitive conclusions could be drawn with regard to the sequential AI-TAM strategy.

Conclusion

Adjuvant AI treatment strategies, including upfront or sequential strategies, are associated with improved cancer recurrences and favourable cost-effectiveness relative to TAM alone approach [iv,v,vi,xxvi,xxvii]. Indirect across-studies comparisons, and early results from BIG 1-98, show that upfront AI strategy is perhaps associated with small incremental QALY gains relative to sequential TAM-AI [xvi,xxix,xxx,xxxii,xxxiii,xxxiv,xxxv,xxxvii]. Not surprisingly, these QALY gains were greater in patients with higher compared to lower recurrence risks [xvi,xxix,xxxiv,xxxv]. Economic evaluations also consistently show that upfront AI is associated with higher incremental costs relative to sequential TAM-AI due to higher upfront drug costs [xxxi,xxxii,xxxiii,xxxiv,xxxv,xxxvii]. Most importantly, the costs per QALY gains (i.e. value for money) seemed favourable in patients who derive higher gains from treatment but unfavourable in those who derive lower gains [xxxiv,xxxv]. A benefit-driven strategy based on predicted magnitude of benefit could, therefore, improve the cost-benefit of AI treatments and help optimize the overall health care system efficiency.

Clinical and economic evaluations to date do not provide a simple answer to the ongoing debate with regard to the optimal adjuvant AI strategy in postmenopausal women with breast cancer. If TEAM or BIG 1-98 [xvi,xvii] confirms that upfront AI is associated with small cumulative QALY gains relative to sequential TAM-AI, as predicted by across-studies comparisons [xxix,xxx,xxxii,xxxiii,xxxiv,xxxv], upfront AI would appear to be a cost-effective strategy in patient who derive higher benefits (e.g. node positive disease or longer life expectancy) while sequential TAM-AI would remain the preferred cost-effective strategy in patients who derive lower benefits (e.g. node negative disease or shorter life expectancy). If BIG 1-98 concludes that sequential AI-TAM is associated with overall outcomes that are comparable to upfront AI with regard to efficacy as well as adverse events and quality of life, it could become the optimal adjuvant AI strategy with economic dominance over upfront AI. In the meantime, rather than choosing one AI strategy for all, a benefit-driven approach according to baseline recurrence risk, and possibly age, is suggested. Further clinical, and economic, research should also attempt to identify optimal AI strategies according to breast cancer associated gene-profile or biomarker signatures.

Acknowledgment

This chapter has been reviewed by Dr Daniel Rayson, Associate Professor of Medicine at Dalhousie University. Halifax. Nova Scotia, Canada.

References

[i] Jemal A, Ward E, Thun MJ. Recent trends in breast cancer incidence rates by age and tumor characteristics among U.S. women. *Breast Cancer Res* 2007; 9(3): R28.

[ii] Carlson RW, Hudis CA, Pritchard KI, et al. Adjuvant endocrine therapy in hormone receptor-positive postmenopausal breast cancer: evolution of NCCN, ASCO, and St Gallen recommendations. *J Natl Compr Canc Netw* 2006; 4(10): 971-979.

[iii] Buzdar AU, Robertson JF, Eiermann W, et al. An overview of the pharmacology and pharmacokinetics of the newer generation aromatase inhibitors anastrozole, letrozole and exemestane. *Cancer* 2002; 95: 2006-2016.

[iv] Rugo HS.The breast cancer continuum in hormone-receptor-positive breast cancer in postmenopausal women: evolving management options focusing on aromatase inhibitors. *Ann Oncol* 2008; 19(1): 16-27.

[v] Perez EA. Safety profiles of tamoxifen and the aromatase inhibitors in adjuvant therapy of hormone-responsive early breast cancer. *Ann Oncol* 2007; 18(Suppl 8): viii26-35.

[vi] Buzdar AU, Coombes RC, Goss PE, Winer EP. Summary of aromatase inhibitor clinical trials in postmenopausal women with early breast cancer. *Cancer* 2008; 112(3 Suppl): 700-709.

[vii] Reeder CA, Gordon D. Managing oncology costs. *Am J Manag Care* 2006; 12(1 Suppl): S3-S16.

[viii] Drummond MF, Sculpher MJ, Torrance GW, O'Brien BJ, Stoddart GL. In: *Methods for the economic evaluation of health care programmes,* 3rd Ed. Oxford: Oxford University Press (2005).

[ix] Briggs A, Claxton K, Sculpher MJ. In: *Decision modeling methods for health economic evaluation.* Oxford: Oxford University Press (2006)

[x] Devlin N, Parkin D. Does NICE have a cost-effectiveness threshold and what other factors influence its decisions? A binary choice analysis. *Health Econ* 2004; 13(5): 437-452.

[xi] Eichler HG, Kong SX, Gerth WC, et al. Use of cost-effectiveness analysis in health-care resource allocation decision-making: how are cost-effectiveness thresholds expected to emerge? *Value Health* 2004; 7(5): 518-528.

[xii] Hirth RA, Chernew ME, Miller E, et al. Willingness to pay for a quality-adjusted life year: in search of a standard. *Med Decis Making* 2000; 20(3): 332-342.

[xiii] Ubel PA, Hirth RA, Chernew ME, Fendrick AM. What is the price of life and why doesn't it increase at the rate of inflation? *Arch Internal Med* 2003; 163: 1637-1641.

[xiv] Forbes JF, Cuzick J, Buzdar A, et al on behalf of The Arimidex, Tamoxifen Alone or in Combination (ATAC) Trialists' Group. Effect of anastrozole and tamoxifen as adjuvant treatment for early stage breast cancer: 100-month analysis of the ATAC trial. *Lancet Oncol* 2008; 9(1): 45-53.

[xv] Coates AKA, Thurlimann B, Mouridsen H, et al. Five years of letrozole compared with tamoxifen as initial adjuvant therapy for postmenopausal women with endocrine-responsive early breast cancer: update of study. *J Clin Oncol* 2007; 25(5): 486-92.

[xvi] Mouridsen HT, Giobbie-Hurder A, Mauriac L, et al on behalf of BIG I-98 Collaborative and the International Breast Cancer Study Group. BIG 1-98: A randomized double-blind phase III study evaluating letrozole and tamoxifen given in sequence as adjuvant endocrine therapy for postmenopausal women with receptor-positive breast cancer [abstract]. *Cancer Res* 2009; 69(Suppl.)(2): 13.

[xvii] Jones SE, Seynaeve C, Hasenburg A, et al. Results of the first planned analysis of the TEAM (tamoxifen exemestane adjuvant multinational) prospective randomized phase III trial in hormone sensitive postmenopausal early breast cancer [abstract]. *Cancer Res* 2009; 69(Suppl.)(2): 15.

[xviii] Ingle JN, Dowsett M, Cuzick J, Davies C on behalf of Early Breast Cancer Trialists' Collaborative Group (EBCTCG). Aromatase inhibitors versus tamoxifen as adjuvant therapy for postmenopausal women with estrogen receptor positive breast cancer: meta-analyses of randomized trials of monotherapy and switching strategies [abstract]. *Cancer Res* 2009; 69(Suppl.)(2): 12.

[xix] Coombes RC, Kilburn LS, Snowdon CF, et al on behalf of the Intergroup Exemestane Study. Survival and safety of exemestane versus tamoxifen after 2–3 years' tamoxifen treatment (Intergroup Exemestane Study): a randomised controlled trial. *Lancet* 2007; 369: 559-570. (Erratum in: Lancet 2007; 369: 906)

[xx] Boccardo F, Rubagotti A, Guglielmini P, et al. Switching to anastrozole versus continued tamoxifen treatment of early breast cancer: preliminary results of the Italian Tamoxifen Anastrozole Trial. *J Clin Oncol* 2005; 23: 5148-5147.

[xxi] Jakesz R, Jonat W, Gnant M, et al on behalf of the ABCSG and the GABG. Switching of postmenopausal women with endocrine responsive early breast cancer to anastrozole after 2 years' adjuvant tamoxifen: combined results of ABCSG trial 8 and ARNO 95 trial. *Lancet* 2005; 366: 455-462.

[xxii] Kaufmann M, Jonat W, Hilfrich J, et al on behalf of the German Adjuvant Breast Cancer Group. Survival benefit of switching to anastrozole after 2 years' treatment with tamoxifen versus continued tamoxifen therapy: the ARNO 95 study. *J Clin Oncol* 2006; 24(18S): 14S (Abstr 547).

[xxiii] Jakesz R, Gnant M, Griel R, et al. Tamoxifen and anastrozole as a sequencing strategy in postmenopausal women with hormone-responsive early breast cancer: updated data from the Austrian breast and colorectal cancer study group trial 8 [abstract]. *Cancer Res* 2009; 69(Suppl.)(2): 14.

[xxiv] Jonat W, Gnant M, Boccardo F, et al. Switching from adjuvant tamoxifen to anastrozole in postmenopausal women with hormone-responsive early breast cancer: a meta-analysis of the ARNO 95 trial, ABCSG trial 8, and the ITA trial. *Breast Cancer Res Treat* 2005; 94: S11 (Abstr 18).

[xxv] Goss PE, Ingle JN, Martino S, et al. Randomized trial of letrozole following tamoxifen as extended adjuvant therapy in receptor-positive breast cancer: updated findings from NCIC CTG MA.17. *J Natl Cancer Inst* 2005; 97: 1262-1271.

[xxvi] Karnon J. Aromatase inhibitors in breast cancer: a review of cost considerations and cost effectiveness. *Pharmacoeconomics.* 2006; 24(3): 215-232.

[xxvii] Annemans L. Methodological issues in evaluating cost effectiveness of adjuvant aromatase inhibitors in early breast cancer: a need for improved modelling to aid decision making. *Pharmacoeconomics.* 2008; 26(5): 409-423.

[xxviii] Early Breast Cancer Trialists' Collaborative Group (EBCTCG). Effects of chemotherapy and hormonal therapy for early breast cancer on recurrence and 15-year survival: an overview of the randomised trials. *Lancet* 2005; 365 (9472): 1687-1717.

[xxix] Punglia RS, Kuntz KM, Winer EP, et al. Optimizing adjuvant endocrine therapy in postmenopausal women with early stage breast cancer: a decision analysis. *J Clin Oncol* 2005; 23: 5178-5187.

[xxx] Cuzick J, Sasieni P, Howell A. Should aromatase inhibitors be used as initial adjuvant treatment or sequenced after tamoxifen? *Br J Cancer* 2006; 94 (4): 460-464.

[xxxi] Lønning PE. Comparing cost/utility of giving an aromatase inhibitoras monotherapy for 5 years vs sequential administration following 2–3 or 5 years of tamoxifen as adjuvant treatment for postmenopausal breast cancer. *Ann Oncol* 2006; 17 (2): 217-25.

[xxxii] Skedgel C, Rayson D, Dewar R, Younis T. Cost-utility of adjuvant hormone therapies for breast cancer in post-menopausal women: sequential tamoxifen-exemestane and upfront anastrozole. *Breast Cancer Res Treat* 2007; 101 (3): 325-333.

[xxxiii] Skedgel C, Rayson D, Dewar R, Younis T. Cost-utility of adjuvant hormone therapies with aromatase inhibitors in post-menopausal women with breast cancer: upfront anastrozole, sequential tamoxifen-exemestane and extended tamoxifen-letrozole. *Breast* 2007; 16 (3): 252-261.

[xxxiv] Younis T, Rayson D, Dewar R, Skedgel C. Modeling for cost effective-adjuvant aromatase inhibitor strategies for post- menopausal women with breast cancer. *Ann Oncol* 2007; 18 (2): 293-298.

[xxxv] Younis T, Rayson D, Skedgel C. T Adjuvant hormonal strategies in postmenopausal women with breast cancer: an economic evaluation based on the EBCTCG meta-analyses [abstract]. *J Clin Oncol* 2009; 27: 15s (suppl; abstr 6574).

[xxxvi] Olivotto IA, Bajdik CD, Ravdin PM, et al. Population-based validation of the prognostic model ADJUVANT! for early breast cancer. *J Clin Oncol* 2005; 23(12): 2716-2725.

[xxxvii] Skedgel C, Rayson D, Younis T. Direct and indirect economic evaluation of upfront and sequential adjuvant treatment in postmenopausal women with breast cancer based on the BIG 1-98 trial [abstract]. *J Clin Oncol* 2009: 27; 15s (suppl; abstr 6594).

In: Aromatase Inhibitors: Types, Mode of Action and Indications ISBN: 978-1-60741-711-8
Editor: Jean R. Lamonte

Chapter V

Aromatase Expression in the Endometrium and its Role in the Development of Uterine Pathology

Hugo Maia Jr, Julio Casoy, Jorge Valente Filho and Elsimar Coutinho

Centro de Pesquisas e Assistência em Reprodução Humana, Salvador, Bahia, Brazil

Abstract

Aromatase expression in the eutopic endometrium is triggered by constant exposure to inflammatory mediators such as prostaglandins, which are produced during menstruation and in the proliferative phase of the menstrual cycle. The presence of aromatase in the endometrial cells is pivotal for the development of endometriosis in the peritoneal cavity, since local estrogen production may be the mechanism that suppresses the phagocytosis of these cells by macrophages and NK cells. The severity of endometriosis correlates positively with the intensity of aromatase expression in the eutopic endometrium, thus corroborating the hypothesis that the progression of endometriosis depends on the constant shedding of aromatase-positive cells through retrograde menstruation. The coexistence of other pathologies with endometriosis is also associated with the presence of aromatase expression in the endometrium. This has been observed in cases of submucous/intramural myomas, endometrial polyps and adenomyosis. In all these cases, the presence of aromatase was associated with increased inflammation in the endometrium as determined by the presence of NF-kappa.B in cell nuclei. The number of NF-kappa.B-positive nuclei was significantly greater in the endometrium of pathology-harboring uteri compared to normal controls. This was observed both during the proliferative phase and prior to the onset of menstruation, thus suggesting a proinflammatory role for estrogens and an antiinflammatory effect of progesterone in the endometrium. Endogenous progesterone or progestins such as gestodene or levonorgestrel when used orally or delivered locally in the uterine cavity are potent inhibitors of aromatase expression in the endometrium and this effect probably involves gene-silencing through transcription inhibition. The decrease in aromatase expression induced by progestins was accompanied by a reduction in the expression of enzymes such as cyclooxygenase type 2 (Cox-2) or angiogenic factors such as VEGF in the endometrium. Inflammation in the endometrium was also diminished by progesterone or progestins and this mechanism involved the inhibition of NF-kappa.B activation. These findings support the role of progestins as anti-aromatase and antiinflammatory agents in the current management of endometriosis and other gynecological pathologies. In fact, the continuous use of combined oral contraceptives containing gestodene or the use of levonorgestrel-releasing intrauterine systems were found to be effective in preventing not only the recurrence of endometriosis but also myoma-related menorrhagia, thereby supporting the causal relationship between aberrant aromatase expression in the endometrium, inflammation and the development of pathology.

Aromatase in the Endometrium and its Regulation by Inflammatory Mediators

Aromatase gene promoter in the endometrium may be activated by prostaglandins and this creates a vicious cycle of inflammation and enhanced local estrogen production [1]. The interplay between inflammation and aromatase expression may be pivotal for the development of pathologies such as endometriosis, adenomyosis, endometrial polyps and myomas, since this enzyme is present in the endometrium of uteri of patients affected by these conditions. However, the triggering stimuli for the activation of this enzyme may lie in the constant exposure of the endometrium to the inflammatory mediators that are activated following progesterone withdrawal prior to the onset of menstruation [2,3]. The fall of progesterone activates the NF-kappa.B proteins, which play an important role in the regulation of proteins that have been linked to cellular transformation, proliferation, apoptosis, angiogenesis and inflammation. In gynecological pathology, NF-Kappa.B is probably one of the major culprits for the development of pathologies such as endometriosis, since all existing and nearly all investigational medications for endometriosis appear to act through suppression of NF-kappa.B activation [4,5].

Aromatase expression in the endometrium, on the other hand, is activated by prostaglandin E2 and since cyclooxygenase II gene transcription is stimulated by NF-Kappa.B, this may establish a link between constant exposure of the endometrium to inflammatory mediators, stimulation of aberrant aromatase and the ensuing development of various gynecological pathologies [3]. This concept will be developed and discussed in the present chapter.

Aromatase Expression in the Endometrium and the Development of Endometriosis

The pathogenesis of endometriosis is not yet completely understood, since controversies remain with respect to several aspects of this disease, hampering the efficacy of treatment in the long term. Clinically, endometriosis behaves as a chronic, recurrent debilitating disease and not as a single pathological event in the reproductive life of a woman that can be resolved with a specific form of treatment. For this reason, it is a pathological condition that may be controlled but never completely cured, since new endometrial cells are shed into the pelvic cavity following successive menstrual periods [6].

The implantation and the development of endometriotic lesions depend on several mechanisms including the repetitive occurrence of retrograde menstruation [6-9]. However, one important issue that remains unresolved is how retrograde menstruation affects the development of endometriosis, since despite its almost universal occurrence in menstruating women it does not result in endometriosis in all cases [6]. It is possible that local factors in the endometrium such as the capacity to synthesize estrogens may play an important role in determining the progression and aggressiveness of endometriotic lesions, which vary considerably from patient to patient [3]. The initiation and development of endometriosis may depend on the extent of aromatase expression in the eutopic endometrium, since there are important differences with respect to gene and enzymatic expression between the eutopic endometrium of these patients and the endometrium of disease-free women [3,6,10]. These endometrial changes may indeed precede the establishment of endometriosis as a clinical, recognizable pathological condition, and they may be responsible for the symptoms associated with this condition.

One important clinical issue in the treatment of endometriosis is whether the currently available therapy is effective in curing the disease in the long range, since most of the meta-analyses carried out so far have indicated high recurrences rates following either surgical treatment or the discontinuation of medical therapy [3,8,10-12]. Nevertheless, rather than any lack of efficacy on the part of the various surgical and medical approaches for the treatment of this disease, these high recurrence rates may reflect more subtle differences between the endometrium of endometriosis and disease-free patients, predisposing the former to recurrence of the disease with the resumption of menstruation [3,9,10]. Aromatase expression in the eutopic endometrium is probably an important risk factor for the development of endometriosis since the clinical behavior of endometriotic lesions may ultimately depend on the amount of aromatase expression that is present in this tissue [1,3]. This hypothesis is in agreement with observations that aromatase expression is not detected in the endometrium of disease-free women, whereas it is present in the endometrium of patients with endometriosis and other pathologies such as adenomyosis, endometrial polyps and myomas [3,13,14].

Aromatase expression in these pathological conditions is enhanced by prostaglandins, particularly prostaglandin E2, and possibly other inflammatory mediators [15]. However, prostaglandin E2 only augments aromatase in the endometrium in which basal levels of this enzyme are detected, having no effect on the aromatase-negative endometrium of disease-free women [2,15,16]. Aromatase expression in the endometrium and its activation by inflammatory mediators such as prostaglandins may, therefore, play a pivotal role in the development of endometriotic lesions in the abdominal cavity [1,9,16]. Retrograde menstruation carries endometrial cells to the pelvis, which upon their arrival there will trigger an intense inflammatory reaction, bringing an efflux of immunologically active cells including macrophages and natural killer cells. However, this intense inflammatory reaction will generate great amounts of prostaglandins and other inflammatory mediators, which will stimulate aromatase expression exponentially in endometrial cells that already have basal levels of aromatase expression, but not in negative cells [1,15]. The rise in aromatase activity will trigger a local production of estrogen, which will further enhance prostaglandin synthesis by activating the cyclooxygenase II enzyme. This will result in a vicious cycle of augmented inflammation and enhanced aromatase expression, and the resulting excessive local estrogen production will facilitate the implantation of endometrial cells in the peritoneum [10, 16]. However, this local hyperestrogenic milieu in the pelvic cavity will only occur if the newly arrived endometrial cells already have basal levels of aromatase expression, since only they have the capacity to respond to inflammation by increasing aromatase expression [15].

The excessive amounts of estrogens produced by these cells will spare them from being destroyed by the efflux of activated macrophage and natural killer cells, since the mechanisms of phagocytosis are disrupted by estrogens [17-19]. One common example of this effect of estrogens is the well-known decrease in bone resorption induced by estrogen during menopause, since osteoclasts are modified macrophages and respond to estrogens by decreasing phagocytosis in the extracellular matrix of the bone. Inhibition of phagocytosis by elevated local estrogen levels in the endometrial cells is probably the mechanism through which these cells in the peritoneal cavity are spared, thus implanting and developing into endometriotic lesions, while the aromatase-negative cells are eliminated from the pelvic cavity in the days following menstruation [15,18,20,21]. This indicates that the clinical behavior of endometriosis may depend on the amount of aromatase expressed in the lesions, which in turn depends on the basal levels of the enzyme present in the eutopic endometrium. Therefore, the presence of an aromatase-positive eutopic endometrium plays a pivotal role in the development of endometriotic lesions and may explain not only why there are such high rates of disease recurrence even after successful medical or surgical treatment but also why some lesions may regress spontaneously without any form of treatment [17,22].

The hypothesis that endometriosis is a chronic condition that depends for its progression on the amount of aromatase expression in the eutopic endometrium predicts that the clinical course of this disease may depend on the amount of inflammation to which the eutopic endometrium is exposed during the menstrual cycle, since the promoter gene for this enzyme responds to prostaglandins [15]. If this hypothesis is correct, it may also explain the high failure rates of all forms of endometriosis treatment in the long range, since the primary defect in the eutopic endometrium was not corrected and may return upon discontinuation of treatment and the resumption of menstruation and its retrograde flow.

Aromatase expression in the eutopic endometrium is, therefore, pivotal not only for the development of endometriosis but also in determining the degree of aggressivity of its clinical course [1,10,23,24]. In this aspect, drugs that suppress aromatase expression or modulate immunological activity will play an important role in endometriosis therapy [25,26]. Recent reports have shown the efficacy of aromatase inhibitors in the treatment of unusually aggressive forms of endometriosis associated with high levels of aromatase expression [23,26,27]. Aromatase expression in the eutopic endometrium is, therefore, crucial in determining the clinical course of this disease including the propensity that some lesions have for spontaneous regression [3]. This can be verified by correlating the aggressiveness of the endometriotic lesion with the presence of aromatase expression in the eutopic endometrium; however, this association may not prove a causal relationship. Previous studies have already shown that the quantity of mRNA transcripts for aromatase were significantly higher in the more aggressive lesions [24].

Recently, our group conducted a clinical observational study to investigate whether more aggressive forms of endometriosis were associated with higher expressions of aromatase expression in the eutopic endometrium. The presence of this enzyme was detected using immunohistochemical methods. Sixty patients with endometriosis who were submitted to laparoscopy had concomitant endometrial biopsies to determine aromatase expression in this tissue. The severity, type and extent of endometriosis were noted and compared with the presence of aromatase expression in the eutopic endometrium. All laparoscopies were carried out during the proliferative phase, since aromatase expression in the endometrium is detected more frequently during this phase than during the luteal phase of the menstrual cycle as previously reported [23]. Initial results have shown that the presence of aromatase expression in the eutopic endometrium of patients showed a positive correlation with the clinical aggressiveness of endometriotic lesions as summarized in Table 1. In patients in whom only fibrotic white lesions of endometriosis were detected at laparoscopy, the frequency of eutopic endometrium that tested positive for aromatase expression was only 19%; however, when active red lesions were present in the peritoneum aromatase expression was detected in 88% of cases. The presence of ovarian endometrioma was also associated with a higher frequency of positive endometrial samples for aromatase expression when compared to fibrotic white lesions. These results, although not establishing a causal relationship between aromatase expression in the eutopic endometrium and the progression of endometriosis, suggest a strong association between the presence of this enzyme in the eutopic endometrium and the detection of active lesions in the pelvis. Regression of endometriosis, on the other hand, was associated with the disappearance of aromatase expression in the eutopic endometrium in the majority of cases. The clinical course of this disease may, therefore, be dictated by the presence of aromatase in the endometrium. All factors that may interfere with aromatase activity in the endometrium or in the endometriotic lesions, including inflammation, may interfere with the clinical course of the disease by either enhancing or decreasing estrogen production in the lesions, thereby affecting the clinical course of the disease [3].

Table 1. Aromatase Expression in the Endometrium in Relation to the Type of Endometriotic Lesion

Red peritoneal lesion	15/17	(88%)*
Black powder burn peritoneal lesion	4/10	(40%)
White fibrotic peritoneal lesion	3/16	(19%)
Ovarian endometrioma	15/17	(88%)

* Significantly higher except for ovarian endometrioma (Chi-square test with significance at $p<0.05$).

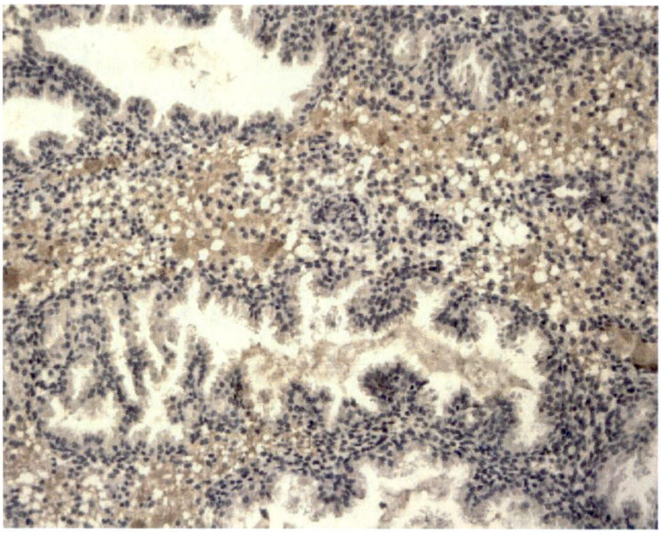

Figure 1. Aromatase positive expression in the endometrial stroma of a patient with red peritoneal lesions of endometriosis.

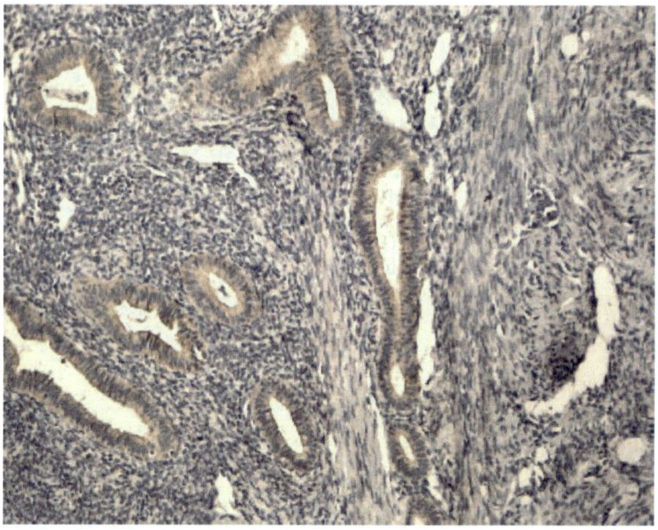

Figure 2. Aromatase expression in the endometrial glands of a patient with adenomyosis.

Aromatase Expression in the Endometrium and its Relationship to Menorrhagia

Several studies have confirmed the capacity of the endometrium to produce estrogens locally, not only in cases of endometriosis but also in the presence of other pathologies such as adenomyosis, endometrial polyps and myomas, which produce clinical symptoms such as menorrhagia and dysmenorrhea [14,23,28-30]. It is speculated that the local production of estrogens in the endometrium may be responsible not only for the development of these pathologies but also for the occurrence of the associated symptoms which may predate their clinical diagnosis [3]. This is in agreement with findings in primates that chronic exposure of the endometrium to estrogens locally in the uterine cavity leads to the development of various pathologies, including adenomyosis, endometrial polyps and myomas, which are associated with aromatase expression in the human endometrium [31].

The aberrant presence of aromatases in the endometrium may upgrade Cox-2 expression in this tissue, thereby resulting in higher prostaglandin levels and further aromatase stimulation, thus creating a positive feedback loop [1,15,28,32] which leads to the exacerbation of symptoms such as pain and uterine bleeding. Epidemiologic studies have shown that patients who later develop endometriosis often report a long history of symptoms before the clinical diagnosis of this pathology is made [8]. This is in agreement with the hypothesis that the development of pathology is the result of the long term exposure of the uterus to the local hyperestrogenism resulting from the augmented endometrial estrogen and prostaglandin E2 synthesis, which initially causes exacerbation of these symptoms and later stimulates the development of these clinical conditions [3].

Adenomyosis is a frequent pathology associated with symptoms such as pain and excessive bleeding during menstruation. In this condition, the basic defect is the anomalous expression of aromatase in the eutopic endometrium, which maintains a positive estrogenic feedback in this tissue, resulting in persistent Cox-2 expression during the luteal phase [14,15]. This will lead not only to increased production of prostaglandins in adenomyosis lesions [33] but also to higher inflammation rates. In the normal uterus, on the other hand, Cox-2 expression becomes negative in most of the endometrial glands during the late luteal phase, although some expression may persist in the superficial epithelium [34]. Aromatase expression in the stroma of the eutopic endometrium of these patients with adenomyosis may, therefore, be the underlying cause of the increased inflammation associated with this and other clinical conditions [14,35].

As inflammatory mediators such as prostaglandins E2 are the triggering stimulus for aromatase expression in adenomyosis and other pathologies, this will create a vicious cycle of increased uterine inflammation and local estrogen production in the endometrium, which may also exert a detrimental effect on fertility [36]. The increased Cox-2 expression in adenomyosis during the luteal phase may play a role in the pathogenesis of dysmenorrhea and menorrhagia associated with this condition, since the intensity of these symptoms correlates positively with levels of prostaglandin production [33].

Table 2. Frequency of Positive Nuclear NF-Kappa.B in the Endometrium (Gland + Stroma) during the menstrual cycle in normal versus pathological uteri

	Normal	Pathology
Proliferative Phase	8/41 (19%)	*24/42 (57%) $p<0.001$
Early Luteal Phase	1/11 (9%)	*3/19 (16%) $p=0.9$(NS)
Late Luteal Phase	20/26 (77%)	*1/7 (14%) $p<0.01$

* Compared to normal. Chi square test with significance at $p<0.005$.

It is known that part of the estrogen-induced effects on uterine growth is mediated by prostaglandins, since they are partially blocked by indomethacin and dexamethasone [37].

The role of prostaglandins and inflammation in augmenting uterine growth may in part explain uterine enlargement in adenomyosis, since Cox-2 and aromatase are expressed continuously in this pathology [14]. Progesterone exerts an antiinflammatory effect on the endometrium, acting through the suppression of NF-kappa B-induced genes. Progesterone withdrawal prior to the onset of menstruation, on the other hand, is associated with the activation of NF-Kappa.B followed by the upregulation of cytokine and prostaglandin production [2,38]. In adenomyosis, it is possible that excessive local estrogen production induced by aberrant aromatase expression mostly in the stroma of the eutopic endometrium may partially counteract the antiinflammatory effects of progesterone, thereby resulting in persistently elevated Cox-2 and activated NF-Kappa.B in both eutopic and ectopic endometrial glands during the luteal phase, whereas in the disease-free endometrium inflammation is kept at a minimum due the anti-inflammatory effect of progesterone [2,14,34].

The increased inflammation resulting from the aberrant aromatase and Cox-2 expression at endometrial level may explain most of the symptoms associated with adenomyosis such as dysmenorrhea, menorrhagia and uterine enlargement [39]. However, it is the translocation of NF-kappa from the cytoplasm to cell nuclei where it binds to DNA that constitutes the initial mechanism through which the inflammatory cascade is activated [2]. Progesterone appears to act by inhibiting this mechanism, since the drop in progesterone levels, which predates the onset of menstrual bleeding, leads to the inactivation of the NF-Kappa-B inhibitory protein in the cytoplasm and the release of active NF-Kappa.b which moves to cell nuclei [2,38]. For this reason, the detection of NF-Kappa by immunohistochemical methods in cell nuclei provides evidence that genes related to inflammation are being activated in that tissue. In the endometrium, the presence of increases in nuclear NF-Kappa during the proliferative and late luteal phases of the menstrual cycle is detected more frequently in patients with uterine pathologies such as adenomyosis than in normal controls (Figure 3). These results are summarized in Table 2. The higher frequency of positive nuclear NF-Kappa.B in uteri with pathologies suggests a positive association between increased inflammation and the development of these clinical conditions. In this context, either estrogenic dominance or progesterone withdrawal plays a pivotal role in the exacerbation of this mechanism [2]. In our study, the uterine pathologies most commonly associated with increased frequency of positive nuclear NF-Kappa.B prior to the onset of menstruation were adenomyosis, polyps and submucous/intramural myomas. It is important to emphasize that it is in these pathologies that aromatase expression is frequently detected either in the endometrium or in the lesions [14,29]. The increased inflammation in adenomyosis may therefore reflect the increased production of estrogens, which alone increases inflammation and prostaglandin production through Cox-2 activation and NF-Kappa.B activation [1-3,15].

Oral contraceptives, on the other hand, because of their strong progestational effect on the endometrium and their suppressive effect on ovarian steroidogenesis, may interrupt this vicious cycle of augmented Cox-2 and aromatase expression at endometrial level, thus decreasing both estrogen synthesis and inflammation. The mechanism of the suppressive effect of oral contraceptives on aromatase expression in the endometrium is direct and it occurs at gene transcriptional level for this enzyme as suggested by our group in a previous publication and confirmed by molecular biology studies [14,40,41]. The lack of detection of the enzyme in the endometrium exposed to progestins by immunohistochemistry is evidence of the impaired formation of this protein due to inhibition of gene transcription [41]. A decrease in inflammation is likewise observed with the use of oral contraceptives. In patients with uterine pathologies and symptoms of menorrhagia and dysmenorrhea, the continuous use of oral contraceptives containing gestodene significantly decreased the frequency of positive nuclear NF-Kappa.B in the endometrium. This effect was observed only in patients in whom amenorrhea was achieved, since the presence of breakthrough bleeding is associated with exacerbation of inflammation (Table 3). These results suggest that the antiinflammatory effects of gestodene on the endometrium are mediated through suppression of NF-Kappa.B activation, thus leading to the inhibition of Cox-2, aromatase and the resolution of pain and bleeding associated with uterine pathologies [40,42].

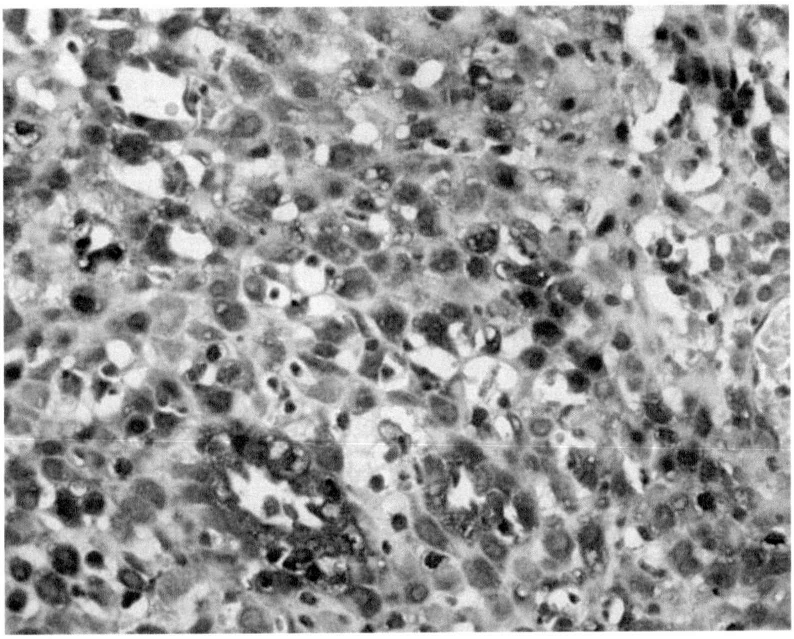

Figure 3. Presence of positive nuclei for NF-Kappa.B in the endometrium of a patient with adenomyosis during the late luteal phase when progesterone levels are declining.

Table 3. Frequency of positive nuclear NF-kappa.B in the deciduas of the endometrium of patients using gestodene 75/ethinyl-estradiol 30 continuously in relation to the presence of uterine bleeding

Breakthrough Bleeding	9/11	(81.8%)*
Amenorrhea	1/30	(3.3%)

$p<0.001$ (Chi-square test).

The effects of gestodene on aromatase are similar to those achieved with danazol, which also inhibits aromatase expression in the eutopic endometrium of patients with adenomyosis by acting directly on gene transcription, thus reducing the levels of messenger RNA [43]. Our findings that oral contraceptives containing gestodene are able to reduce aromatase expression in the eutopic endometrium of patients with adenomyosis, endometriosis and myomas may explain their effectiveness when used to prevent not only recurrences in surgically treated endometriosis patients but also the bleeding and pain associated with these pathologies [40,42,44]. In patients with adenomyosis and other associated pathologies, the use of a levonorgestrel-releasing intrauterine device also successfully suppressed aromatase expression in the endometrium (Maia H Jr. Unpublished results). This may explain its effectiveness in treating menorrhagia caused by adenomyosis [45].

In conclusion, the detection of aromatase activity in basal glands in the eutopic endometrium of patients with adenomyosis suggests a pivotal role for the basalis in the pathogenesis of adenomyosis [6]. The hyperestrogenic milieu caused by aberrant aromatase expression may be important not only for the mechanism of invagination of the basalis into the myometrium, but also for the exacerbation of inflammation associated with this pathology.

Aromatase Expression and Myomas

In the endometrium of patients with submucous/intramural myomas, aromatase expression is detected more frequently during the proliferative phase than in the luteal phase of the menstrual cycle [23,35]. The presence of aromatase expression in the endometrium of patients with myomas shows, as expected, an association with the presence of abnormal uterine bleeding. In patients with subserous myomas but without menorrhagia, aromatase expression was absent in the endometrium, while in patients with submucous myomas, aromatase expression was frequently detected [35]. This is in agreement with the hypothesis that aromatase expression plays a causative role in the development of symptoms such as abnormal uterine bleeding or pain, which are frequently associated with myomas. The increase in local estrogen production in the endometrium of patients with myomatous uteri will exacerbate the inflammation and NF-Kappa.B activation, as previously described in this chapter.. In fact, an increase in prostaglandin production and inflammation was observed in uteri with submucous and intramural myomas but not in those with subserous myomas. This corroborates the hypothesis that increased local estrogen production will enhance inflammation and create a vicious cycle that will ultimately lead to abnormal uterine bleeding. Although this is inconclusive with respect to whether the presence of aromatase expression in the endometrium may account for the presence of abnormal menstrual bleeding in patients with myomatous uteri, aberrant aromatase expression has been detected in the presence of other uterine pathologies associated with excessive uterine bleeding and pain such as adenomyosis, endometriosis and endometrial polyps [14,29,46].

The frequent association of symptomatic myomas with endometriosis, adenomyosis and endometrial polyps also suggests that these pathologies share the same pathogenesis. There is also evidence that local estrogen production in myomas or in the eutopic endometrium is more important than systemic estrogens with respect to growth stimulation, since myoma growth cannot be fully restored by exogenous estrogens in patients using GnRH analogs because of the suppressive effects of GnRH analogs on aromatase expression in myomatous uteri [47]. These findings are in agreement with the hypothesis that locally produced estrogens acting in an intracrine fashion either in the endometrium or in the myoma may play a pivotal role in both the pathogenesis of abnormal menstrual bleeding and in the growth of this neoplasm [23,35].

However, the rate and velocity of fibroid growth may involve other factors besides estrogenic stimuli. In cases of subserous myomas, which may attain very large sizes without provoking abnormal uterine bleeding, the rapid growth rate is more dependable on cytogenetic rearrangements than on hormonal factors, since aromatase expression is negative in this neoplasm [29,48-50]. In addition to stimulating inflammation, the local estrogens will stimulate the synthesis of angiogenic factors such as VEGF in the endometrium, ultimately leading to the menorrhagia associated with the presence of myomas [23,51]. It is possible that endometrial alterations such as increased inflammation and vascularization resulting from endometrial estrogen production may be the cause of myoma-related menorrhagia rather than the tumor itself. This is in agreement with the finding that menorrhagia is positively related to the presence of aromatase expression in the endometrium rather than to myoma size [35]. It would be an oversimplification to consider this local estrogen production the only pathogenic mechanism for myoma development, since other factors may act synergistically, including cytogenetic abnormalities that may affect growth rates irrespective of the intensity of symptoms [52-54].

The use of GnRH analogs or danazol is associated not only with reduction of bleeding in patients with myomas but also with the inhibition of aromatase expression in the endometrium [46]. Another benefit of the use of progestins is the reduction in the relative risk of developing myomas in women using oral contraceptives, which, likewise, may be a consequence of the inhibition of aromatase expression in the endometrium of uteri containing myomas [23,35]. The suppression of aromatase p450 expression in the endometrium will also ultimately reduce the formation of angiogenic factors such as VEGF [23]. The blockade of inflammation and angiogenesis may, therefore, be the prime mechanism by which oral contraceptives are effective not only in inhibiting the growth rate of myomas but also in reducing the amount of uterine bleeding associated with the presence of these tumors [55,56].

Aromatase Expression in the Endometrium of Patients with Unexplained Infertility

It has recently been reported that the presence of aromatase decreases endometrial receptivity to implantation in *in vitro* fertilization cycles [57]. Aromatase expression in the endometrium at the critical time of implantation may disturb the delicate balance between estrogen and progesterone, thereby affecting the mechanism of decidualization and implantation. The presence of aromatase induces a shift to a hyperestrogenic endometrial milieu that impedes the full decidualization of the stromal fibroblasts in response to progesterone. This probably constitutes the mechanism underlying the implantation failure associated with the presence of high endometrial levels of mRNA for aromatase [57,58]. Interference with implantation due to a shift in the balance between estrogen and progesterone resulting from aberrant aromatase expression in the endometrium has been suggested as a possible mechanism for the reduction of fertility observed in patients with minimal endometriosis in which the distortion of pelvic anatomy has not yet occurred [3]. In fact, aromatase expression in the endometrium may be the initial step in reducing fertility by impairing implantation and this probably precedes the development of pathologies such as endometriosis, adenomyosis and endometrial polyps [3,29,40].

In patients with three or more repeated IVF failures in which pretreatment hysteroscopy revealed a normal uterine cavity, a second procedure carried out after the repeated failures revealed the appearance of intrauterine pathologies including endometrial polyps in 18% of the cases [59]. In these patients, it is possible that the underlying defect, possibly the aberrant aromatase expression, was already present in the endometrium at the time of initial hysteroscopy prior to IVF, and was responsible for the implantation failures despite the fact that the pathology had not yet developed. The posterior detection of the intrauterine pathology in the second hysteroscopy in some of these patients may reflect the time course of this process in which the prolonged exposure of the endometrium to local estrogens first impairs implantation and decreases fertility to later favor the development of pathology. Unfortunately, the presence of aromatase in the endometrium was not investigated in this study. However, in patients with endometrial polyps, aromatase was detected not only in the lesion but also in the adjacent endometrium, indicating that the primary defect lies in this tissue [3,29,60].

Because of the paucity of studies on the presence of aromatase in the endometrium of patients with either unexplained infertility or repeated IVF failures, it is impossible to ascertain the role of this enzyme in human infertility. Recently, our group investigated the presence of aromatase expression in the endometrium of patients with either unexplained infertility or repeated IVF failures. Thirty-six patients who had normal findings at transvaginal sonography, hysteroscopy and laparoscopy were enrolled for this observational study. Because of the diagnosis of unexplained infertility and the longstanding infertility, they were offered cycles of IVF which proved unsuccessful. In this group of patients with IVF implantation failures, the positive rate for aromatase in the endometrium was 80%. In normal patients, on the other hand, aromatase expression is not detected in the endometrium. As these patients have not yet been diagnosed with any pathology, this is in agreement with the hypothesis that aromatase expression in the endometrium induced by constant exposure to prostaglandins during menstruation may not only be responsible for the decreased fertility but may also place these patients at greater risk for the development of pathologies, including endometrial polyps, adenomyosis, myomas and endometriosis [3].

Drugs that inhibit aromatase expression in the endometrium without suppressing ovulation should be the ideal medication to treat endometriosis-related infertility. This is particularly important since available medical therapies for endometriosis, albeit effective for the treatment of pain, do not improve pregnancy rates when compared to placebo. This may be explained by the fact that although progestins are able to suppress aromatase expression in the endometrium [14,41], they also suppress ovulation, thereby preventing the occurrence of pregnancy. The discontinuation of medication, on the other hand, leads to resumption of ovulation and menstruation, which may expose the endometrium to prostaglandins and the subsequent activation of aromatase. Danazol, which has been extensively used to treat endometriosis, is a drug that suppresses both ovulation and aromatase expression at gene transcription level [43]. However, when danazol was used vaginally instead of orally, the blood levels achieved were insufficient to block ovulation. Initial trials using danazol at a dose of 1500 mg in a vaginal ring to be replaced every two months showed that it was indeed possible to treat endometriosis-related pain without blocking ovulation. In addition, significantly fewer side effects were reported compared to those observed when the oral route was used. Most importantly, since ovulation was not blocked, over half the patients became pregnant while using the danazol-loaded vaginal ring with no untoward effects on the ongoing pregnancy [61]. Although the effects of the vaginal ring containing danazol on aromatase expression in the endometrium were not investigated in this preliminary clinical study, it is possible that the beneficial effects of this treatment on both fertility and endometriosis-related pain may be the result of a direct inhibitory effect of danazol on aromatase expression [46].

The use of the vaginal route to deliver other hormones that effectively suppress aromatase expression in the endometrium when given orally, but that do not have the disadvantage of suppressing ovulation, may prove to represent a good therapeutic approach for the treatment of endometriosis-related infertility. The normalization of aromatase expression may explain the pain-reducing effects of danazol-containing vaginal rings in patients with endometriosis of the ovary and vaginal septum. This is in agreement with the clinical findings that levonorgestrel-releasing intrauterine devices are effective for the treatment of endometriosis-related pain without suppressing ovulation, since they inhibit aromatase expression in the endometrium. The report of the occurrence of ongoing pregnancies in patients in use of the danazol-bearing vaginal ring is also in agreement with the hypothesis that there is a local effect on the endometrium that suppresses aromatase expression. It is possible that vaginal rings containing danazol may be suitable for the treatment of unexpected infertility; however, this hypothesis has yet to be confirmed.

Most importantly, in patients undergoing IVF treatment, the use of danazol prior to the initiation of ovarian stimulation almost doubled the rate of clinical pregnancies, and this effect was attributed to better endometrial receptivity achieved with the prior use of danazol [62]. Unfortunately, whether this fertility enhancing effect of danazol was due to aromatase inhibition at endometrial level was not part of the protocol of that study. However, since danazol is a potent inhibitor of aromatase activity in the endometrium and the presence of this enzyme lowers implantation rates, it is logical to assume that the fertility-enhancing effects of this drug prior to IVF may be due to aromatase inhibition. Many cases of unexplained infertility are caused by defects of implantation and possibly in a great majority of cases this may be a consequence of aberrant aromatase expression. The drawback of using oral danazol is that ovulation is inhibited during drug administration, which prevents the occurrence of pregnancy during treatment. Since this does not occur with the use of the vaginal ring, it is possible that this route of administration of danazol may be more effective in the treatment of unexplained infertility compared to the oral route of administration. This is certainly an area that deserves further investigation. However, further studies are still necessary to affirm whether aromatase expression in the endometrium has a causative role in unexplained infertility. The silencing of the expression of this enzyme in the endometrium without the concomitant blockade of ovulation by using vaginally administered progestins, on the other hand, may in the future become one of the effective ways of treating both unexplained and endometriosis-related infertility.

Aromatase Expression in the Postmenopausal Endometrium and its Relation to Pathology in Tibolone Users

The role of inflammation and aromatase expression in endometrial pathology in menopausal patients using tibolone has yet to be investigated. In an open prospective cohort study comparing tibolone with conventional hormone replacement therapy (HRT), raloxifene and controls, the incidence of bleeding and spotting in tibolone users was 20%, which represented an almost 10-fold increase in relation to controls and raloxifene users, although it was still half the incidence observed in conventional HRT users [63].

The use of hysteroscopy to evaluate the uterine cavity in patients using tibolone and experiencing uterine bleeding revealed that the risk of developing endometrial polyps was greater in tibolone users compared to users of other forms of HRT [64,65]. However, the mechanisms by which tibolone increases the risk of polyp formation in the endometrium remain to be fully explained. It has recently been proposed that the main mechanism of action of tibolone at tissue level is not through its metabolites, but rather from the increase in free testosterone resulting from the decreased levels of sex hormone binding globulin (SHBG) in circulation [66]. Because endometrial polyps express aromatase, a plausible explanation for the development of this neoplasm in tibolone users would be the increased intratumor conversion of free testosterone into estradiol. The atrophic endometrium, on the other hand, lacks detectable aromatase activity and therefore is unable to convert testosterone into estradiol.

In an attempt to clarify the mechanisms by which tibolone affects polyp development, our research group has investigated the presence of aromatase expression in the endometrium of tibolone users who were experiencing abnormal uterine bleeding during the use of this therapy. Aromatase was detected by immunohistochemical methods in endometrial samples obtained by hysteroscopic endometrial resection. The study was observational and was carried out in 23 menopausal patients who were experiencing abnormal uterine bleeding at the time of hysteroscopy. The pathological diagnosis was adenomyosis and atrophic endometrium in 8 cases, endometrial polyps in 6 cases and atrophic endometrium in 9 patients. In the atrophic endometrium, aromatase expression was not detected. However, when adenomyosis was present, aromatase expression was detected in the glandular epithelium of the eutopic endometrium and adenomyosis in 60% of cases. In patients with endometrial polyps, aromatase expression was detected in the neoplasm but not in the surrounding atrophic endometrium in 50% of the cases. It is important to note the large proportion of cases of adenomyosis and endometrial polyps in which aberrant aromatase expression was detected in endometrial tissue or in the neoplasm.

Tibolone is devoid of any effect on the normal atrophic endometrium; however, it can stimulate the development of preexisting pathologies such as endometrial polyps and adenomyosis, since aromatase expression is present in these cases. The decreased levels of SHBG in circulation and the resulting increase in free testosterone, readily converted to estrogens by the action of tissue aromatase, offer the perfect scenario for the development of pathologies that already express this enzyme. This hypothesis would explain the higher risk of developing endometrial polyps in tibolone users when compared with the estrogen-progestin group and also the reactivation of dormant adenomyosis. The presence of aromatase in endometrial polyps and adenomyosis will increase local estrogen production and hasten the development of these pathologies through the local conversion of the increased levels of free testosterone to estrogens [14,67].

Conclusion

The presence of aromatase expression in the endometrium is a risk factor for the development of pathology, since the local production of estrogens will stimulate inflammation. This mechanism is complex and involves a vicious cycle of augmented production of prostaglandin E2 through Cox-2 activation in response to the increased levels of estrogens. The occurrence of subclinical inflammation in the endometrium in the presence of various estrogen-dependent pathologies can also be inferred by the existence of activated NF-Kappa.B expression in cell nuclei, an important transcriptional factor in the inflammatory cascade. The activation of nuclear NF-Kappa.B is stimulated by estrogen exposure or by progesterone withdrawal, since the constant exposure of the endometrium to progestins suppresses inflammation and inhibits NF-Kappa.B translocation to cell nuclei. Progestins are therefore important antiinflammatory hormones whose actions on the endometrium are capable of suppressing NF-kappa activation and reducing the expression of enzymes such as aromatase and Cox-2. These effects are of paramount importance for their clinical application in the treatment of pathologies caused by chronic inflammation and aberrant aromatase expression such as endometriosis. The augmented local estrogen production catalyzed by the presence of aromatase in the endometrium by increasing inflammation and the ensuing excessive prostaglandin formation is also responsible for the exacerbation of menstrual-related symptoms such as dysmenorrhea and menorrhagia.

References

[1] Attar E. Aromatase and other steroidogenic genes in endometriosis. Translational aspects. *Human Reproduction Update* 2006; 12:49-56.

[2] Modugno F, Ness RB, Chen C *et al.* Inflammation and endometrial cancer: a hypothesis. *Cancer Epidemiol Biomarkers Prev* 2005; 14: 2840-7.

[3] Maia H Jr, Casoy J, Valente J. Is aromatase expression in the endometrium the cause of endometriosis and its related infertility? *Gynecol.Endocrinol* 2009. In Press.

[4] Guo SW. Nuclear factor-kappa.B (NF-kappaB): an unsuspected major culprit in the pathogenesis of endometriosis that is still at large? *Gynecol Obstet Invest.* 2007;63(2):71-97.

[5] Sugino N, Karube-Harada A, Taketani T, Sakata A, Nakamura Y. Withdrawal of ovarian steroids stimulates prostaglandin F2alpha production through nuclear factor-kappaB activation via oxygen radicals in human endometrial stromal cells: potential relevance to menstruation. *J Reprod Dev.* 2004 Apr;50(2):215-25.

[6] Leyendecker G, Herbertz M, Kunz G, Mall G. Endometriosis results from dislocation of basal endometrium. *Hum Reprod* 2003; 18: 1130-1.

[7] Metzger DA, Szpak CA, Haney AF. Histologic features associated with hormonal responsiveness of ectopic endometrium. *Fertil Steril* 1993; 59:83-88.

[8] Hadfield R, Mardon H, Barlow DH, Kennedy S. Delay in diagnosis of endometriosis: a survey of women from the USA and the UK. *Human Reprod.* 1996; 11:878-80.

[9] Kitawaki J, Kado N, Ishihara H, Koshiba H, Kitaoka Y, Honjo H. Endometriosis: the pathophysiology as estrogen dependent diseases. *J. Steroid Biochem. Mol. Biol.* 2002; 83:149-55.

[10] Ulukus M, Cakmak H, Arici A. The role of endometrium in endometriosis. *J Soc Gynecol. Investig* 2006; 13(7):467-76.

[11] Exacoustos C, Zupi E, Amadio A, Amoroso C, Szabolcs B, Romanini ME, Arduini D. Recurrence of endometriomas after laparoscopic removal: sonographic and clinical follow-up and indication for second surgery. *J Minim Invasive Gynecol.* 2006 Jul-Aug; 13(4): 281-8.

[12] Nieto A, Jacuri C, Serra M, Keller J, Cortes-Prieto J. Long term follow-up of endometriosis after two different therapies (gestrinone and buserelin). *Clin. Exp Obstet Gynecol.* 1996; 23:198-204.

[13] Velasco I, Rueda J, Acien P. Aromatase expression in endometriotic tissues and cell cultures of patients with endometriosis. *Mol Hum Reprod.* 2006 Jun; 12(6):377-81.

[14] Maia H Jr, Casoy J, Correia T, Freitas A, Pimentel K, Coutinho EM. Effect of the menstrual cycle and oral contraceptives on aromatase and Cox-2 expression in adenomyosis. *Gynecological Endocrinology* 2006; 22:1-5.

[15] Noble LS. Prostaglandin E2 stimulates aromatase activity in endometriosis derived stromal cells. *Journal Clinical Endocrinology Metabolism* 1997; 82:600-606.

[16] Bulun SE, Zeitoun K, Takayana K, Noble L et al. Oestrogen production in endometriosis and use of aromatase inhibitors to treat endometriosis. *Endocrine Rel. Cancer* 1999; 6:293-301.

[17] Olive DL, Montoya I, Riehl RM, Schenken RS. Macrophage conditioned media enhance endometrial stroma proliferation in vitro. *Am J Obstet Gynecol* 1991; 164(4): 953-8.

[18] Sidell N, Han SW, Parthasarathy S. Regulation and modulation of abnormal immune responses in endometriosis. *Ann N Y Acad Sci.* 2002; 955:159-73.

[19] Carlstein H. Immune responses and bone loss: the estrogen connection. *Immunol Rev.* 2005; Dec 2008: 194-206. Review.

[20] Lebovic DI, Muller MD, Taylor RN. Immunobiology of endometriosis. *Fertil Steril* 2001; 75:1.

[21] Capellino S, Montagna P, Villaggio B, Sulli A, Soldano S, Ferrero S, Remorgida V, Cutolo M. Role of estrogens in inflammatory response: expression of estrogen receptors in peritoneal fluid macrophages from endometriosis. *Ann N Y Acad Sci.* 2006 Jun; 1069:263-67.

[22] Jones KD, Haines P, Sutton CJ. Long term follow-up of controlled trial of laser laparoscopy for pelvic pain. *JSLS* 2001; 5:111-5.

[23] Maia H Jr, Casoy J, Pimentel K, Correia T, Athayde C, Cruz T, Coutinho EM. Effect of oral contraceptives on vascular endothelial growth factor, Cox-2 and aromatase

expression in endometrium of uteri affected by myomas and associated pathologies. *Contraception* 2008 78(6):479-85.

[24] Acien P, Velasco I, Gutierrez M, Martinez-Beltran M. Aromatase expression in endometriotic tissues and its relationship to clinical and analytical findings. *Fertil Steril.* 2007;88(1):32-8.

[25] D'Hooghe TM. Immunomodulators and aromatase inhibitors: are they the next generation of treatment for endometriosis? *Curr Opin Obstet Gynecol* 2003; 15:243-9.

[26] Fatemi HM, Al-Turki HA, Papanikolaou EG, Kosmas L, De Sutter P, Devroey P. Successful treatment of an aggressive recurrent post-menopausal endometriosis with an aromatase inhibitor. *Reproductive BioMedicine Online,* Volume 11, Number 4, October 2005, pp. 455-457(3).

[27] Amsterdam LL, Gentry W, Jobanputra S, Wolf M, Rubin SD, Bulun SE. Anastrazole and oral contraceptives: a novel treatment for endometriosis. *Fertil Steril.* 2005; 84 (2):300-4.

[28] Bulun SE, Zeitoun KM, Takayama K, Sasano H. Molecular basis for treating endometriosis with aromatase inhibitors. *Human Reproduction Update* 2000;6:413-18

[29] Kitawaki J, Kusuki I, Koshiba H, Tsukamoto K, Honjo H. Expression of aromatase cytochrome p450 in eutopic endometrium and its application as a diagnostic test for endometriosis. *Gynecol and Obstet Invest* 1999; 48 (suppl 1) 21-28.

[30] Urabe M, Yamamoto T, Kitawaki J, Honjo H, Okada H. Estrogen biosynthesis in human uterine adenomyosis. *Acta Endocrinol (Copenh)* 1989; 121:259-64.

[31] Baskin GB, Smith SM, Marx PA. Endometrial hyperplasia, polyps and adenomyosis associated with unopposed estrogens in rhesus monkey (Macaca mulata). *Vet Pathol* 2002; 39:572-75.

[32] Zeitoun K, Bulun SE. Aromatase: a key molecule in the pathophysiology of endometriosis and a therapeutic target. *Fertil & Steril* 1999; 72:961-69.

[33] Koike H, Egawa W, Ohtsuka T, Yamaguchi M, Ikenoue T, Mori N. Correlation between dysmenorrheic severity and prostaglandin production in women with endometriosis. *Prostaglandins Leukot Essent Fatty Acids* 1992; 46:133-37.

[34] Maia H Jr, Maltez A, Studard E, Zausner B, Athayde C, Coutinho EM. Effect of menstrual cycle and oral contraceptives on cycloooxygenase-2 expression in the endometrium. *Gynecol Endocrinol* 2005; 21:57-61.

[35] Maia H Jr, Pimentel K, Casoy J, Correia T, Freitas LA, Zausner B, Athayde C, Coutinho EM. Aromatase expression in the eutopic endometrium of myomatous uteri: the influence of menstrual cycle and oral contraceptives. *Gynecol Endocrinol* 2007 Jun;23(6):320-4.

[36] Shimizu Y, Fukuda J, Kumasawa Y, Tanaka T. Pregnancy complicated by adenomyosis resulted in miscarriage in three cases of in-vitro fertilization - embryo transfer. *Reprod. Med Biol* 2004; 3: 95-98.

[37] Stewart PJ, Zaloudeck CT, Imman MM, Webster RA. Effect of dexamethasone and indomethacin on estrogen induced uterine growth. *Life Sci* 1983; 33:2349-56.

[38] Critchley HO, Kelly RW, Brenne RM, Baird DT. The endocrinology of menstruation. A role for the immune system. *Clin Endocrinol* 2001; 55: 701-10.

[39] Ferenczy A. Pathophysiology of adenomyosis. Hum Reprod Update 1998; 4:312-322.

[40] Maia H Jr. and Casoy J. Non-contraceptive health benefits of oral contraceptives .*Eur J Contracept Reprod Health Care 2008 Mar* 13(1):17-24.

[41] Fechner S, Husen B, Thole H, Schmidt M, Gashaw I, Kimmig R, Winterhager E, Grummer R. Expression and regulation of estrogen-converting enzymes in ectopic human endometrial tissue. *Fertil Steril* 2007; 88(4 Suppl):1029-38.

[42] Fraser I S, Kovacs GT. The efficacy of non-contraceptive uses for hormonal contraceptives. *MJA* 2003; 178:621-3.

[43] Ishihara H, Kitawaki J, Kado N, Koshiba H, Fushiki S, Honjo H. Gonadotrophin-releasing hormone agonist and danazol normalize aromatase cytochrome p450 expression in eutopic endometrium from women with endometriosis, adenomyosis or leiomyomas. *Fertil Steril* 2003, 79 suppl 1:735-42.

[44] Maia Jr. H, Coelho G, Valente Filho J, Athayde C, Barbosa IC, Coutinho EM. Prevention of endometriotic cyst recurrence with continuous oral contraceptives. *Journal of Contraception and Reproductive Health* 2004; v9:p152.

[45] Maia Jr H, Maltez A, Coelho G *et al*. Insertion of Mirena after endometrial resection in patients with adenomyosis. *J Am Assoc Gynecol Laparosc* 2003; 10: 512-6.

[46] Maia H Jr ,Casoy J, Correia T, Freitas LA, Pimentel K, Athayde C. The effect of oral contraceptives on aromatase expression in the eutopic endometrium of patients with endometriosis. *Gynecologic Endocrinol* 2008 24(3);123-8.

[47] Hozu M, Sumitani H et al. Overexpression of aromatase P-450 in leiomyoma tissue is driven through the promoter 1.4 of aromatase p-450. *J Clin Endocrinol Metab* 87: 2540-48, 2002.

[48] Brosens I, Deprest J, Dal Cin P, Van den Berghe H. Clinical significance of cytogenetic abnormalities in uterine myomas. *Fertil.Steril.* 1998;69:232-5.

[49] Bulun SE et al. Mechanisms of excessive estrogen formation in endometriosis. *Journal of Reproductive Immunology* 2002;55:21-33.

[50] Ebert AD et al. Aromatase inhibitors and cyclooxygenase-2 (Cox-2) inhibitors in endometriosis. New questions old answers? *European Journal Obstetrics Gynecology & Reproductive Biology* 2005;122:144-50.

[51] Sugino N, Kashida S, Karube-Harada A, Takiquchi S, Kato H. Expression of Vascular Endothelial Growth Factor and its receptor in human endometrium throughout the menstrual cycle and early pregnancy. *Reproduction* 2002;123:379-89.

[52] Otsuka H, Shinohara M et al. A comparative study of the estrogen receptor ratio in myometrium and uterine leiomyomas. *Int.J Gynecol Obstet.* 1989;29:189-94.

[53] Kataoka S, Yamada H, Hoshi N, Kudo M, Hareyama H, Sakuragi N, Fugimoto S. Cytogenetic analysis of uterine leiomyoma: the size, histopathology and GnRH a response in relation to chromosome karyotype. *Eur J Obstet Gyecol Reprod Biol* 2003;110(1):58-62.

[54] Crow J. Pathology of uterine fibroids. Baillieres Clin Obstet Gynecol 1998;12:197-211.

[55] Orsini G, Laricchia L, Fanelli M. Low dose combination oral contraceptive use in women with uterine leiomyomas. *Minerva Ginecol* 2002;54:253-61.

[56] Jensen JT, Speroff L. Health Benefits of Oral Contraceptives. *Obstet Gynecol Clinic of North America* 2000;27:705-22.

[57] Brosens J, Verhoeven H, Campo R, Gianaroli L, Gordts S, Hazekamp J, Hägglund L, Mardesic T, Varila E, Zech J, Brosens I. High endometrial aromatase P450 mRNA expression is associated with poor IVF outcome. *Hum Reprod.* 2004 Feb;19(2):352-6

[58] Aghajanova L, Hamilton A, Kwintkiewicz J, Vo KC, Giudice LC. Steroidogenic enzyme and key decidualization marker dysregulation in endometrial stromal cells from women with versus without endometriosis. *Biol Reprod.* 2009 Jan;80(1):105-14.

[59] Dicker D, Ashkenazi J, Feldberg D, Farhi J, Shalev J, Ben-Rafael Z. The value of repeat hysteroscopic evaluation in patients with failed in vitro fertilization transfer cycles. *Fertil Steril.* 1992 Oct;58(4):833-5.

[60] Pal L, Niklaus AL, Kim M, Pollack S, Santoro N. Heterogeneity in endometrial expression of aromatase in polyp-bearing uteri. *Hum Reprod.* 2008 Jan;23(1):80-4. Epub 2007 Nov 7.

[61] Igarashi M, Izuka M, Abe Y, Ibulki Y. Novel vaginal danazol ring therapy for pelvic endometriosis in patients with deeply infiltrating endometriosis. *Hum Reprod* 1998 13(7) ;1952-6.

[62] Tei C, Mivazaki T et al. Effect of danazol on pregnancy rate in patients with in vitro fertilization-embryo transfer. *J. Repro. Med* 1998; 43:541-6.

[63] Christodoulakos GE, Botsis DS, Lambrinoudaki IV, Papagianni VD, Panoulis CP, Creatsa MG, Alexandrou AP, Augoulea AD, Dendrinos SG, Creatsas GC. A 5-year study on the effect of hormone therapy, tibolone and raloxifene on vaginal bleeding and endometrial thickness. *Maturitas.* 2006 Mar 20;53(4):413-23.

[64] Perez-Medina T, Bajo-Arenas J, Haya J, Sanfrutos L, Iniesta S, Bueno B, Castelo-Branco C. Tibolone and risk of endometrial polyps: a prospective, comparative study with hormone therapy. *Menopause.* 2003 Nov-Dec;10(6):534-7.

[65] Ginsburg J, Prelevic GM. Cause of vaginal bleeding in postmenopausal women taking tibolone. *Maturitas.* 1996 May;24(1-2):107-10.

[66] Maia H Jr,, Casoy J, Valente J. Testosterone replacement therapy in menopause. Benefits beyond sexuality. *Gynecologic Endocrinol* 2009 25(1) 12-20.

[67] Maia H Jr, Pimentel K, Silva TM, Freitas LA, Zausner B, Athayde C, Coutinho EM. Aromatase and cox-2 expression in endometrial polyps during the menstrual cycle. *Gynecol Endocrinol.* 2006 Apr;22(4):219-24.

In: Aromatase Inhibitors: Types, Mode of Action and Indications ISBN: 978-1-60741-711-8
Editor: Jean R. Lamonte

Chapter VI

Aromatase Inhibitors for the Medical Treatment of Endometriosis

Kaei Nasu**, Akitoshi Tsuno, Akitoshi Yuge, Masakazu Nishida and Hisashi Narahara
Department of Obstetrics and Gynecology, Faculty of Medicine, Oita University, Oita, Japan

Abstract

Endometriosis, a common, benign, estrogen-dependent disease affecting 3-10% of women of reproductive age, is characterized by the ectopic growth of endometrial tissue found primarily in the peritoneum, ovaries, and rectovaginal septum. Women suffering from endometriosis may present with chronic pelvic pain, dysmenorrhea, dyspareunia, and/or subfertility. The prevalence of endometriosis in women with pelvic pain and/or subfertility is estimated to be between 20-90%; thus, it is one of the most frequently encountered benign gynecological problems.

Various surgical and medical strategies have been used to treat endometriosis. Medical therapies historically have included contraceptive steroids, progestogens, and agonists of gonadotropin-releasing hormone, as well as androgens and non-steroidal anti-inflammatory agents. Current medical treatment aims to inhibit the growth of endometriotic implants by suppressing ovarian steroids and inducing a hypoestrogenic state. However, treatments for endometriosis that aim to lower circulating estradiol concentrations can be used for only a limited time due to unacceptable side effects. In addition, high recurrence rates of up to 45% after the completion of medical treatments

* Address for correspondence: Kaei Nasu, MD, PhD, Department of Obstetrics and Gynecology, Faculty of Medicine, Oita University, Idaigaoka 1-1, Hasama-machi, Yufu-shi, Oita 879-5593, Japan. Tel: 81 97 586 5922; Fax: 81 97 586 6687; E-mail: nasu@med.oita-u.ac.jp

remain a significant problem. Therefore, novel therapeutic strategies are necessary to improve the clinical management of patients with endometriosis.

Aromatase P450 is a key enzyme for estrogen biosynthesis, because aromatase catalyzes the conversion of androstenedione and testosterone to estrone and estradiol. Increasing evidence suggests that endometrial aromatase expression is involved in the pathogenesis of endometriosis. Recent studies show that aromatase expression is consistently found in endometriotic lesions and in the eutopic endometrium from patients with endometriosis, whereas it is absent in eutopic endometrium from women without endometriosis. The aromatase enzyme, which is demonstrated to be present locally in endometriotic implants, is indicative of estrogen biosynthesis in these lesions. Local estrogen production by these implants could contribute to the progression of endometriosis even during treatment with GnRH agonists, which inhibit only ovarian production of estrogen. A non-steroidal aromatase inhibitor is a highly potent and specific suppressor of estrogen production locally and systemically. Several reports have demonstrated the therapeutic role of the aromatase inhibitor in patients with endometriosis.

This paper is a review of the recent literature concerning basic and clinical research on the use of aromatase inhibitors to medically treat endometriosis. The mechanisms of action, efficacy, and adverse effects of aromatase inhibitors in the treatment of endometriosis are discussed.

Keywords: endometriosis; aromatase; third-generation aromatase inhibitor; medical treatment

Introduction

Endometriosis, a common, benign, estrogen-dependent disease affecting 3-10% of women of reproductive age and up to 50% of premenopausal women, is characterized by the ectopic growth of functional endometrial tissue outside the uterus; such tissue is found primarily in the pelvic peritoneum, ovaries, and rectovaginal septum [Wheeler, 1989; Vessey et al., 1993; Olive and Schwartz, 1993; Zeitoun et al., 1999; Bulun et al., 2000; Bulun et al., 2002]. Women suffering from endometriosis may present with chronic pelvic pain, dysmenorrhea, dyspareunia, and/or subfertility [American College of Gynecologists, 1999; Giudice and Kao, 2004]. The prevalence of endometriosis in women with pelvic pain and/or subfertility is estimated to be between 20-90% [Koninckx et al., 1991; Carter, 1994]. Thus it is one of the most frequently encountered benign gynecological problems that require hospitalization, and a leading cause of hysterectomy [Gazvani and Templeton, 2002]. Moreover, there is an association between untreated endometriosis and development of ovarian cancer [Brinton et al., 1997].

Symptoms such as chronic pelvic pain, dyspareunia, dysmenorrhea and painful bowel movements are believed to arise from cyclical bleeding into the surrounding tissues, resulting in inflammation and the formation of scars and adhesions. The development of endometriosis depends on estrogen, and several features of this disease can be explained by the overproduction of estrogen [Zeitoun et al., 1998]. Various surgical and medical strategies have been used to treat endometriosis. Medical therapies historically have included contraceptive steroids in the cyclic or continuous mode, progestogens, and agonists of gonadotropin-releasing hormone (GnRH), as well as androgens and non-steroidal anti-inflammatory agents [Olive and Pritts, 2001; Practice Committee of the American Society for Reproductive Medicine, 2004; Giudice and Kao, 2004; Olive et al., 2004]. Standard medical treatments aim to inhibit the growth of endometriotic implants by either inducing hypoestrogenism or antagonizing estrogen action [Practice Committee of the American Society for Reproductive Medicine, 2004]. However, almost all of these treatment modalities failed to treat the endometriosis-associated pain. These medical treatments for endometriosis can be used only for a limited time due to their unacceptable side effects, and high recurrence rates up to 45% after the completion of medical treatment have been the most significant problem [Bergqvist, 2000; Appleyard et al., 2006]. One report indicates that 9% of women are initially resistant to treatments, and 50% of the responders have recurrence of pain at the end of a 6-month follow-up [Vercellini et al., 1997]. The recurrence rate of pelvic pain within 5 years after "medical oophorectomy" induced by GnRH agonists in severe endometriosis remains as high as 74% [Henzl et al., 1988; Waller and Shaw, 1993]. Therefore, novel therapeutic strategies are necessary to improve the clinical management of patients with endometriosis.

Because of the systemic hypoestrogenic symptoms associated with currently available hormonal therapies that block ovarian function, new approaches are being studied for endometriosis that have the potential to target the estrogen-dependent growth of endometrial lesions more selectively [Chwalisz et al., 2002; Viganò et al., 2003]. These include aromatase inhibitors, selective estrogen receptor modulators, and selective progesterone receptor modulators. Aromatase inhibitors in combination with an oral contraceptive (OC) or progestin were found to be effective in patients with endometriosis who were resistant to these existing medical and surgical treatments [Ailawadi et al., 2004; Amsterdam et al., 2005]. In a placebo-controlled prospective randomized study, an aromatase inhibitor in combination with a GnRH agonist regimen showed a strikingly superior response compared with a GnRH agonist-only regimen with respect to the recurrence of pain after 24 months of therapy [Soysal et al., 2004]. Therefore, the novel treatment with aromatase inhibitors may have significant clinical advantages over the currently used treatments with respect to pain relief and the recurrence of pain in patients with endometriosis. The clinical significance of estrogen biosynthesis in endometriosis is exemplified by clinical observations that estrogen is essential for the growth of endometriosis. Previous reports demonstrated abundant aromatase expression and local estrogen production in endometriotic tissue [Kitawaki et al., 1997; Zeitoun et al., 1999; Bulun et al., 2001; Fang et al., 2002; Gurates et al., 2002; Yang et al., 2002]. The subsequent introduction of aromatase inhibitors into the treatment of endometriosis successfully underscored the presence of aromatase in endometriotic tissue [Takayama et al., 1998; Ailawadi et al., 2004]. These recent results suggested that aromatase inhibitors might treat endometriosis more effectively than GnRH analogues by suppressing local estrogen formation in endometriotic tissue.

We review herein the published clinical evidence for the use of aromatase inhibitors in the treatment of endometriosis. The action mechanisms, efficacy, and adverse effects of aromatase inhibitor treatment for endometriosis are also discussed.

Aromatase-mediated Estrogen Biosynthesis

The biologically active estrogen 17β-estradiol is produced from cholesterol through six serial enzymatic conversions in two ovarian cell types that co-operate in a paracrine fashion. The two rate-limiting steps include the entry of cholesterol into the mitochondrion, facilitated by the steroidogenic acute regulatory (StAR) protein and the conversion of androstenedione to estrone by aromatase. Aromatization (i.e., the conversion of C19 steroids to estrogens) is a very important step in estrogen formation [Gruber et al., 2002]. Aromatase is a cytochrome P450 enzyme that catalyzes the final and rate-limiting step in estrogen biosynthesis, the conversion of androgens to estrogens [Buzdar, 1999]. The CYP19 gene encodes a steroid aromatase that catalyzes the conversion of C-19 androgens to estrogens. Androstenedione and testosterone serve as substrates for aromatase; through steroid hydroxylation, these androgens are converted to estrone and 17β-estradiol (Figure 1). Aromatase enzyme complex is comprised of two polypeptides. The first of these is a specific cytochrome, P450, the product of the CYP19 gene [Simpson et al., 2002]. The second is a flavoprotein, nicotinamide-adenine dinucleotide phosphate-cytochrome P450 reductase, which is distributed ubiquitously in most cells.

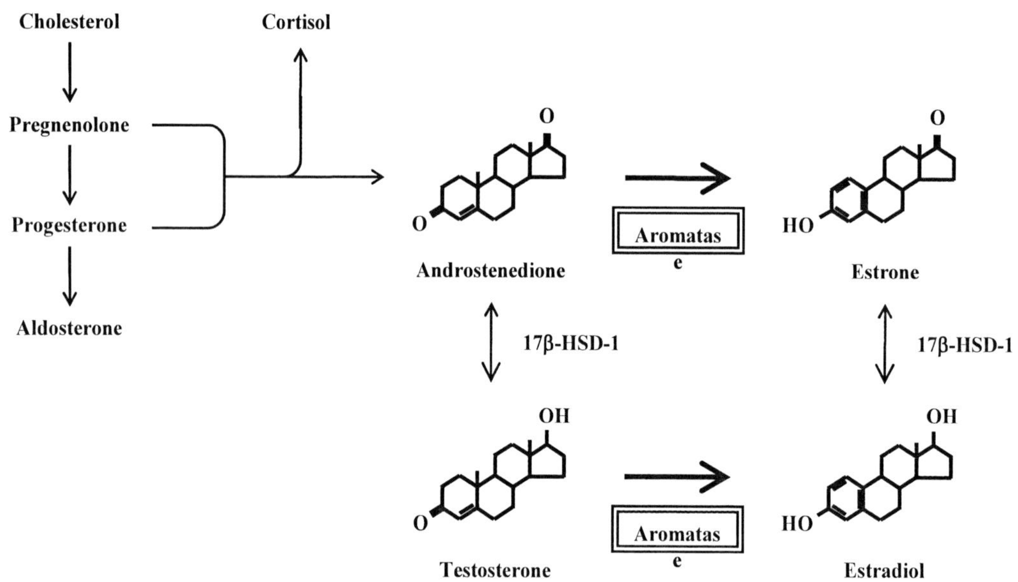

Figure 1. Steroidogenic pathway. Estradiol is produced from cholesterol by a number of steroidogenic enzymes. Aromatase mediates the final and rate-limiting step in estrogen biosynthesis, the conversion of androgens to estrogens.

The aromatase enzyme is localized in the endoplasmic reticulum of estrogen-producing cells [Sebastian et al., 2002; Simpson et al., 2002]. Aromatase is expressed in a number of human tissues and cells, such as ovarian granulosa cells, placental syncytiotrophoblast, and testicular Leydig cells, as well as in various extraglandular sites including adipose tissue and skin fibroblasts, and in the brain [Simpson et al., 1994]. Therefore, estrogen synthesis by aromatase occurs not only in the ovary but also in a number of tissues throughout the body.

Recent studies indicate that aromatase gene transcription is highly regulated [Simpson et al., 1997]. Aromatase catalyzes estrogen formation in several human tissues under the control of alternatively used promoters. Transcription of the aromatase gene in human tissues is regulated by at least 10 distinct promoters. In the reproductive-age woman, the ovary is the most important site of estrogen biosynthesis, and this takes place in a cyclic fashion. Upon binding of follicle-stimulating hormone (FSH) to its G-protein-coupled receptor in the granulosa cell membrane, intracellular cyclic adenosine monophosphate (cAMP) levels rise and enhance the binding of two critical transcription factors steroidogenic factor-1 (SF-1) and cAMP response element binding protein (CREB) to the classically located proximal promoter II of the aromatase gene [Michael et al., 1995; Michael et al., 1997]. This, in turn, activates aromatase expression and consequently estrogen secretion from the pre-ovulatory follicle [Simpson et al., 1994; Michael et al., 1995].

Aberrant Aromatase Expression in Endometriosis

Endometriotic tissue growth increases with the estrogen level. Endometrial implants contain estrogen, progesterone, and androgen receptors [Prentice et al. 1992; Bergqvist and Fernö 1993], as well as aromatase, an enzyme that catalyses the conversion of androgens to estrogens. Ectopic endometriotic implants respond to ovarian hormonal changes. Local estrogen production in conjunction with circulating estrogen stimulates the growth of endometriosis [Kitawaki et al. 1997; Bulun et al., 1999; Bulun et al., 2000].

Aromatase expression has been demonstrated in the stromal cells of endometriotic tissues [Kitawaki et al., 1997; Noble et al., 1997; Zeitoun and Bulun, 1999; Bulun et al., 2001; Tsai et al., 2001; Sun et al., 2003], which apparently provide a local source of estrogens within the pathologic tissue itself [Zeitoun and Bulun, 1999]. Abundant aromatase enzyme activity is also present in stromal cells isolated and cultured from ovarian endometriotic cysts [Noble et al., 1997; Zeitoun et al., 1999; Tsai et al., 2001; Sun et al., 2003]. In addition to aromatase, StAR and other necessary steroidogenic enzymes are expressed in vivo in endometriosis, enabling this tissue to synthesize estradiol from cholesterol de novo [Zeitoun et al., 1999; Bulun et al., 2001; Tsai et al., 2001]. The expression of the enzyme that inactivates estradiol by conversion to estrone, 17β-hydroxysteroid dehydrogenase, is low in endometriotic tissue [Bulun et al., 2000]. Importantly, the product of aromatase activity in these tissues, namely estrone, is only very weakly estrogenic, and must therefore be converted to estradiol-17β in tissue sites of action. Prostaglandin (PG) E2 or a cAMP analog stimulates the expression of aromatase, StAR, and other steroidogenic genes in endometriotic stromal cells in a time- and concentration-dependent fashion [Tsai et al., 2001]. Prostaglandin E2 stimulates the production of progesterone, 17-hydroxyprogesterone, androstenedione, and estradiol primarily via the induction of aromatase in endometriotic cells [Noble et al., 1997; Tsai et al., 2001]. Therefore, the secretion of circulating estradiol not only by the ovary, adipose tissue, and skin but also locally by endometriotic tissue itself may contribute greatly to the formation of endometriotic implants [Noble et al., 1996; Noble et al., 1997].

Aromatase expression or steroid hormone production is virtually undetectable in eutopic endometrial stromal cells from disease-free women [Noble et al., 1997; Bulun et al., 2004].

Aromatase activity or mRNA could not be induced by PGE2 or cAMP analogs in stromal cells from disease-free women [Noble et al., 1997]. On the other hand, low but significant levels of aromatase enzyme activity and mRNA were detected in eutopic endometrial tissue and stromal cells of women with endometriosis [Noble et al., 1996; Noble et al., 1997]. This is consistent with the observations of many other investigators that the eutopic endometrial tissue of women with endometriosis exhibits multiple molecular abnormalities that may be linked to aberrant production of estrogen, prostaglandins, and cytokines, or to decreased production of molecules important for differentiation and implantation [Lessey et al., 1994; Kao et al., 2003; Wu et al., 2005].

Third-generation Aromatase Inhibitors

Aromatase is an excellent target for the inhibition of estradiol synthesis, because it is the last step in steroid biosynthesis, there are no important downstream enzymes to be affected. In addition, although aromatase is a P450 enzyme and shares common features with other enzymes in this class, such as liver-metabolizing and steroidogenic enzymes, it has unique features for the aromatizing reaction, making it amenable to selective inhibition [Brodie and Long, 2001]. Aromatase inhibitors inhibit the conversion of adrenal androgens into estrogens in peripheral adipose tissue and, to a lesser extent, in muscle, which is the main source of estrogens in postmenopausal women. Aromatase inhibitor use in postmenopausal women decreases the level of circulating estrogen to 1-10% of the pretreatment level [Winer et al., 2002].

The aromatase inhibitors are classified as either type I (suicidal or noncompetitive) or type II (competitive) inhibitors [Buzdar and Howell, 2001; Goss and Strasser, 2001]. Both types compete for binding to the active site. Once a type I inhibitor has bound, the enzyme initiates hydroxylation, which produces an unbreakable bond between the inhibitor and the enzyme protein. Enzyme activity thus is permanently blocked [Buzdar et al., 2002]. Exemestane (Aromasin; Pfizer, New London, CT) is a type I (suicidal) inhibitor. Type II inhibitors reversibly bind to the active enzyme site, and no enzyme activity is triggered. The inhibitor can disassociate from the binding site, allowing renewed competition between the inhibitor and the substrate for binding to the site. Thus, continued activity requires the constant presence of an inhibitor, and the effectiveness of competitive inhibitors depends on the affinities of the inhibitor and the substrate [Buzdar et al., 2002]. Anastrozole (Arimidex; AstraZeneca, Wilmington, DE) [Plourde et al., 1994] and letrozole (Femara; Novartis, East Hanover, NJ) [Bhatnagar et al., 1990] are type II inhibitors (Figure 2). Anastrozole, letrozole, and exemestane are classified as third-generation aromatase inhibitors and are highly potent and specific suppressors of estrogen production locally and systemically. The third-generation agents are associated with fewer side effects, and are sufficiently long-acting to be administered on a once-a-day basis.

Aromatase inhibitors were developed to act on sex steroid–dependent neoplasms by suppressing in situ estrogen production [De Jong et al., 1997]. Two third-generation competitive inhibitors (anastrozole and letrozole) and one inactivator (exemestane) have now been approved for use in the United States and worldwide [Plourde et al., 1994; Buzdar et al.,

1997; Buzdar and Hortobagyi, 1998; Scott and Wiseman, 1999; Bhatnagar et al., 2001; Goss and Strasser, 2001; Winer et al., 2002; Goldhirsch et al., 2005]. Two triazole compounds, anastrozole and letrozole, are highly potent, competitive, and reversible aromatase inhibitors. These compounds lower estrogen levels in patients to the limit of assay sensitivity and inhibit peripheral aromatization in postmenopausal breast cancer patients. The third-generation aromatase inhibitors have shown better efficacy in terms of disease-free survival compared to tamoxifen as an initial adjuvant hormonal therapy, or compared to when aromatase inhibitors were given sequentially after tamoxifen therapy [Goss et al., 2003; Coombes et al., 2004; ATAC Trialists' Group, 2005]. They are now used in the adjuvant setting and as first-line treatment in the metastatic disease.

In premenopausal women, the ovaries can overcome an estrogen blockade by reflex increments of luteinizing hormone (LH) and FSH, so aromatase inhibitors become inactive when used alone in premenopausal women [Santen et al., 1980]. Because of insufficient ovarian blockade in premenopausal women with breast cancer, aromatase inhibitors must be combined with a GnRH agonist to prevent the reflex LH and FSH increments [Stein et al., 1990; Carlson, 2002; Forward et al., 2004].

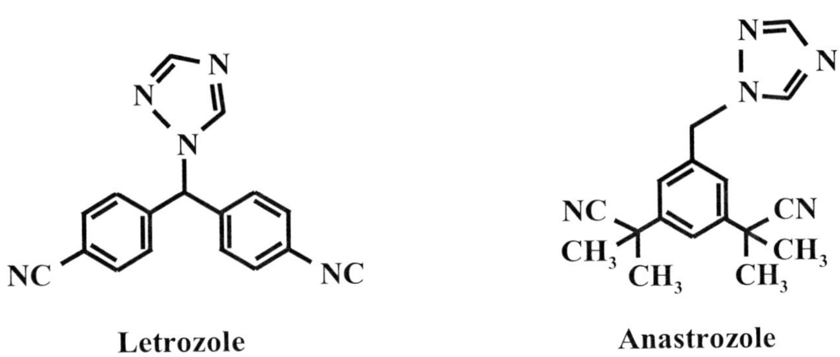

Figure 2. The chemical structure of third generation non-steroidal aromatase inhibitors, anastrozole and letrozole.

Side Effects of Aromatase Inhibitors

In general, aromatase inhibitors are well tolerated and are not associated with severe complications [ATAC Trialists' Group, 2006; Jenkins et al., 2008]. Their most common short-term adverse effects include mood changes, fatigue, nausea, diarrhea, back pain, arthralgia, myalgia, dyspnea, headache, hot flushes, vaginal dryness, and breakthrough bleeding [Baum et al., 2003; ATAC Trialists' Group, 2006; Sestak et al., 2008]. However, these adverse effects are mild and infrequent. Very rarely, letrozole and anastrozole can cause a skin rash. The majority of these side effects were mild and improved as treatment progressed.

Long-term use of aromatase inhibitors carries the potential risk of osteopenia, osteoporosis, and/or bone fracture [Buzdar, 2004; ATAC Trialists' Group, 2006; Sestak et al., 2008]. Results from ATAC trials after completion of adjuvant treatment of breast cancer for 5 years showed that the safety profile of anastrozole remains unchanged during the entire treatment period [Buzdar, 2004; ATAC Trialists' Group, 2006]. The overall fracture rate during 5 years of anastrozole treatment (7.1%) was significantly higher than that associated with tamoxifen (4.6%) treatment. The anastrozole plus OC pilot trial in the treatment of premenopausal patients with endometriosis showed no significant changes in bone density during 6 months of treatment [Amsterdam et al., 2005]. On the other hand, a randomized controlled clinical trial using anastrozole plus the GnRH agonist showed significant bone loss after 6 months of treatment [Soysal et al., 2004]. The observed bone mineral density loss was significantly greater at 6 months in the GnRH agonist plus anastrozole arm vs. the GnRH agonist-only arm. This effect persisted even after treatment ceased. However, none of these patients became osteopenic or osteoporotic. The regimens that combine aromatase inhibitors with add-back progestins or OCs do not appear to be associated with significant bone loss after 6 months of treatment and therefore may be suitable for long-term use. Administration of calcium citrate and vitamin D is recommended together with aromatase inhibitors and careful bone mineral density assessment [Ailawadi et al., 2004]. Adverse effects of aromatase inhibitors on bone are predictable and can be treated, or even prevented in at-risk patients, with bisphosphonates [Coleman, 2003]. Alternatively, estrogen add-back therapy could be utilized to protect bones [Chwalisz et al., 2002].

Aromatase Inhibitors for the Treatment of Endometriosis

Because of estrogen's importance in stimulating endometriotic tissues and the in situ presence of aromatase in these tissues, the inhibition of estrogen synthesis is a rational approach to treatment [Zeitoun and Bulun, 1999; Bulun et al., 1999; Bulun et al., 2000]. Aromatase inhibitors inhibit proliferation and induce apoptosis of eutopic endometrial epithelial cells in patients with endometriosis [Meresman et al., 2005]. The recently introduced highly specific aromatase inhibitors have successfully treated pelvic pain and significantly reduced lesion size in both postmenopausal and premenopausal women [Takayama et al., 1998; Ailawadi et al., 2004; Razzi et al., 2004; Shippen and West, 2004; Soysal et al., 2004; Amsterdam et al., 2005; Attar and Bulun, 2006]. The ability to convert androgen precursors endogenously into estrogens by endometriotic cells is probably the major reason for the failure of GnRH agonists, which reduce only ovarian estrogens [Bulun et al., 1999]. Aromatase inhibitors are now used as adjunctive therapy to GnRH agonist, OC, and progestogens in women with resistant endometriosis-related symptoms [Takayama et al., 1998; Ailawadi et al., 2004]. Table 1 summarizes these clinical studies of aromatase inhibitor use in endometriosis [Takayama et al., 1998; Ailawadi et al., 2004; Razzi et al., 2004; Shippen and West, 2004; Soysal et al., 2004; Amsterdam et al., 2005; Fatemi et al., 2005; Hefler et al., 2005; Remorgida et al., 2007; Bohrer et al., 2008].

Takayama et al. [1998] reported the first case of postmenopausal endometriosis treated with an aromatase inhibitor, anastrozole. Circulating levels of estradiol-17β were reduced to approximately 50% of the baseline value after the onset of treatment with anastrozole. An aromatase inhibitor effectively eradicated the tumor and eliminated her pain. This was followed by another case report confirming that the use of an aromatase inhibitor is the medical treatment of choice for postmenopausal endometriosis [Razzi et al., 2004]. Aromatase inhibitors would provide an effective solution in those rare cases of endometriosis in postmenopausal women with scarce ovarian estradiol production. However, these clinical effects need confirmation by a prospective powerful randomized controlled trial.

Table 1. Clinical reports of the use of aromatase inhibitors in patients with endometriosis

Author (Year)	Study design	Menopausal status	No. of Patients	Intervention	Treatment length (mo)	Outcome
Takayama et al. (1998)	Case report	Postmeno-pausal	1	Anastrozole	9	Pain relief, reduced lesion size
Ailawadi et al. (2004)	Pilot prospective	Premeno-pausal	10	Letrozole + progestogen	6	Pain relief, reduced lesion size
Razzi et al. (2004)	Case report	Postmeno-pausal	1	Letrozole	9	Pain relief, reduced lesion size
Shippen and west (2004)	Case report	Premeno-pausal	2	Letrozole + progestogen	6	Pain relief, reduced lesion size
Soysal et al. (2004)	Randomized controlled trial	Premeno-pausal	80	Anastrozole + GnRH agonist	6	Pain relief
Amsterdam et al. (2005)	Pilot prospective	Premeno-pausal	10	Anastrozole + OC	6	Pain relief
Fatemi et al. (2005)	Case report	Postmeno-pausal	1	Letrozole	18	Pain relief, reduced lesion size
Hefler et al. (2005)	Pilot prospective	Premeno-pausal	10	Vaginal anastrozole	6	Better QOL
Remorgida et al. (2007)	Pilot prospective	Premeno-pausal	12	Letrozole + progestogen	2-4	Pain relief, ovarian cysts
Bohrer et al. (2008)	Case report	Postmenopausal	1	Anastrozole	15	Ineffective for ureteral obstruction

Information regarding the use of aromatase inhibitors in women of reproductive age is still limited [Shippen and West, 2004; Ailawadi et al., 2004; Amsterdam et al., 2005; Hefler et al., 2005]. By blocking the conversion of androgens to estrogens in ovarian granulosa cells, these drugs reduce negative feedback at the pituitary–hypothalamus level, and therefore increase serum FSH levels that stimulate the development of ovarian follicles [Mitwally and Casper, 2004]. Thus, when aromatase inhibitors are administered to premenopausal women for the treatment of endometriosis, they must be combined with a gonadotropin-blocking agent such as GnRH agonist, progestins, or OCs in order to prevent the development of functional ovarian cysts [Shippen and West, 2004; Ailawadi et al., 2004; Soysal et al., 2004; Amsterdam et al., 2005].

Soysal et al. [2004] performed a randomized placebo controlled trial to assess the clinical efficacy of anastrozole in conjunction with goserelin, as compared with goserelin alone. Eighty patients with severe endometriosis according to the American Society for Reproductive Medicine (ASRM) criteria were enrolled into the study after conservative surgery. The first group of patients received anastrozole plus depot injections of goserelin every 4 weeks, and the second group of patients received a placebo tablet in addition to the goserelin regimen for 24 weeks. The GnRH agonist plus aromatase inhibitor regimen showed a more profound, stable, and long-lasting effect on chronic pain (54.7% vs. 10%) with minimal adverse effects.

A small phase II trial and a case report, in which an aromatase inhibitor was administered together with a progestogen, have been published. In the pilot trial, 10 premenopausal patients resistant to existing medical and surgical treatments for endometriosis were administered an aromatase inhibitor, letrozole, and norethindrone acetate daily for 6 months [Ailawadi et al., 2004]. Endometriosis was evaluated by pre- and post-treatment laparoscopies. Both pelvic pain scores and ASRM laparoscopic scores decreased significantly. Nine of 10 patients responded to this regimen with decreased pelvic pain. In a report by Shippen and West [2004], an aromatase inhibitor, anastrozole, together with oral progestogen was given for 6 months to 2 premenopausal women with endometriosis. Both were diagnosed by laparoscopy and failed to respond to OCs, GnRH agonist, and nonsteroidal anti-inflammatory and analgesic medications. Treatment resulted in a rapid, progressive reduction in symptoms. Pregnancy was achieved in both cases. Bone densitometries were in the mid-normal ranges 18 months after completion of treatment. Side effects were minimal and well tolerated by both patients. On the other hand, it has been reported that the combination of letrozole and progestin caused the formation of multiple ovarian cysts in the majority of patients [Ailawadi et al., 2004; Amsterdam et al., 2005; Hefler et al., 2005].

A prospective open-label pilot study employing a novel regimen consisting of an aromatase inhibitor and combination OC for premenopausal endometriosis has been reported [Amsterdam et al., 2005]. In that study, each patient was started on a daily regimen of anastrozole plus an OC for 6 months. A significant reduction in pain was noted in 14 of 15 patients at the end of the study. Pain scores fell starting at month 1 and continued to decrease with each subsequent month of treatment.

Conclusions

As discussed in this review, third-generation aromatase inhibitors are candidate drugs for the treatment of endometriosis, which is resistant to existing therapeutic modalities. The aromatase inhibitors appear to be effective in alleviating pain, reducing lesion size, and probably improving life quality associated with endometriosis. In postmenopausal patients, aromatase inhibitors can be used alone without severe adverse effects. In premenopausal women with endometriosis, on the other hand, the use of the aromatase inhibitors requires ovarian suppression via the addition of a GnRH agonist, progestogens, or combination OC. It is supposed that the combination of an aromatase inhibitor with a progestogen or OC will gain more popularity over that of an aromatase inhibitor with a GnRH agonist, because the former are simpler, cheaper, associated with fewer side effects, and may be administered for longer periods. Alternatively, lower doses of aromatase inhibitors also potentially may be used in the treatment of pain or infertility associated with endometriosis. Patients with endometriosis are usually at the reproductive stage and may still want to have a child [Santen, 2003; Soysal et al., 2004]. More randomized clinical trial data are needed to confirm the findings from these studies and to look at the efficacy and safety of these regimens combining aromatase inhibitors with gonadotropin-blocking agents [Patwardhan et al., 2008]. The question of teratogenicity or embryotoxicity should be further evaluated, especially if the therapeutic application of aromatase inhibitors is considered for young patients with endometriosis. The goal of future research should be to define the optimal roles of the different medical options in the treatment of endometriosis.

Acknowledgments

This work was supported in part by a Grant-in-Aid for Scientific Research from the Japan Society for the Promotion of Science (no. 20591920 to K. Nasu).

References

American College of Gynecologists. (1999) Medical management of endometriosis: clinical management guidelines for obstetrician-gynecologists. *ACOG Prac Bull* 11: 978-990.

Amsterdam LL, Gentry W, Jobanputra S, Wolf M, Rubin SD, Bulun SE. (2005) Anastrozole and oral contraceptives: a novel treatment for endometriosis. *Fertil Steril* 84: 300-304.

Appleyard T, Mann C, Khan K. (2006) Guidelines for the management of pelvic pain associated with endometriosis: a systematic appraisal of their quality. *Br J Obstet Gynaecol* 113: 749-757.

ATAC Trialists' Group. (2006) Comprehensive side-effect profile of anastrozole and tamoxifen as adjuvant treatment for early-stage breast cancer: long-term safety analysis of the ATAC trial. *Lancet Oncol* 7: 633-643.

Attar E, Bulun SE. (2006) Aromatase inhibitors: the next generation of therapeutics for endometriosis? *Fertil Steril* 85: 1307-1318.

Ailawadi RK, Jobanputra S, Kataria M, Gurates B, Bulun SE. (2004) Treatment of endometriosis and chronic pelvic pain with letrozole and norethindrone acetate: a pilot study. *Fertil Steril* 81: 290-296.

Baum M, Buzdar A, Cuzick J, Forbes J, Houghton J, Howell A, Sahmoud T; The ATAC (Arimidex, Tamoxifen Alone or in Combination) Trialists' Group. (2003) Anastrozole alone or in combination with Tamoxifen versus Tamoxifen alone for adjuvant treatment of postmenopausal women with early-stage breast cancer: results of the ATAC (Arimidex, Tamoxifen Alone or in Combination) trial efficacy and safety update analyses. *Cancer* 98: 1802-1810.

Bergqvist A. (2000) A comparative study of the acceptability and effect of goserelin and nafarelin on endometriosis. *Gynecol Endocrinol* 14: 425-432.

Bergqvist A, Fernö M. (1993) Oestrogen and progesterone receptors in endometriotic tissue and endometrium: comparison of different cycle phases and ages. *Hum Reprod* 8: 2211-2217

Bhatnagar AS, Brodie AMH. Long BJ, Evans DB, Miller WR. (2001) Intracellular aromatase and its relevance to the pharmacological efficacy of aromatase inhibitors. *J Steroid Biochem Mol Biol* 76: 199-202.

Bhatnagar AS, Hausler A, Trunet P, Schieweck K, Lang M, Bowman P. (1990) Highly selective inhibition of estrogen biosynthesis by CGS 20267, a new non-steroidal aromatase inhibitor. *J Steroid Biochem Mol Biol* 37: 1021-1027.

Bohrer J, Chen CC, Falcone T. (2008) Persistent bilateral ureteral obstruction secondary to endometriosis despite treatment with an aromatase inhibitor. 90: 2004.e7-2004.e9.

Brinton LA, Gridley G, Persson I, Baron J, Bergqvist A. (1997) Cancer risk after a hospital discharge diagnosis of endometriosis. *Am J Obstet Gynecol* 176: 572-579.

Bulun SE, Fang Z, Imir G, Gurates B, Tamura M, Yilmaz B, Langoi D, Amin S, Yang S, Deb S. (2004) Aromatase and endometriosis. *Semin Reprod Med* 22: 45-50.

Bulun SE, Gurates B, Fang Z, Tamura M, Sebastian S, Zhou J, Amin S, Yang S. (2002) Mechanisms of excessive estrogen formation in endometriosis. *J Reprod Immunol* 55: 21-33.

Bulun SE, Yang S, Fang Z, Gurates B, Tamura M, Zhou J, Sebastian S. (2001) Role of aromatase in endometrial disease. *J Steroid Biochem Mol Biol* 79: 19-25.

Bulun SE, Zeitoun K, Takayama K, Noble L, Michael D, Simpson E, Johns A, Putman M, Sasano H. (1999) Estrogen production in endometriosis and use of aromatase inhibitors to treat endometriosis. *Endocrine Related Cancer* 6: 293-301.

Bulun SE, Zeitoun K, Takayama K, Sasano H. (2000) Molecular basis for treating endometriosis with aromatase inhibitors. *Hum Reprod Update* 6: 413-418.

Buzdar AU. (1999) Role of aromatase inhibitors in advanced breast cancer. *Endocrine Related Cancer* 6: 219-225.

Buzdar AU. (2004) The ATAC (Arimidex, Tamoxifen, Alone or in Combination) trial: an update. *Clin Breast Cancer 5 Suppl* 1: S6-S12.

Buzdar AU, Hortobagyi G. (1998) Update on endocrine therapy for breast cancer. *Clin Cancer Res* 4: 527-534.

Buzdar AU, Howell A. (2001) Advances in aromatase inhibition: clinical efficacy and tolerability in the treatment of breast cancer. *Clin Cancer Res* 7: 2620-2635.

Buzdar AU, Jonat W, Howell A, Plourde PV. (1997) ARIMIDEX: a potent and selective aromatase inhibitor for the treatment of advanced breast cancer. *J Steroid Biochem Mol Biol* 61: 145-149.

Buzdar AU, Robertson JF, Eiermann W, Nabholtz JM. (2002) An overview of the pharmacology and pharmacokinetics of the newer generation aromatase inhibitors anastrozole, letrozole, and exemestane. *Cancer* 95: 2006-2016.

Brodie A, Long B. (2001) Aromatase inhibition and inactivation. *Clin Cancer Res* 7: 4343s-4349s.

Carlson RW. (2002) Sequencing of endocrine therapies in breast cancer – integration of recent data. *Breast Cancer Res Treat* 75: S27-S32.

Carter JE. (1994) Combined hysteroscopic and laparoscopic findings in patients with chronic pelvic pain. *J Am Assoc Gynecol Laparosc* 2: 43-47.

Chwalisz K, Garg R, Brenner RM, Schubert G, Elger W. (2002) Selective progesterone receptor modulators (SPRMs): a novel therapeutic concept in endometriosis. *Ann N Y Acad Sci* 955: 373-388.

Coleman RE. (2003) Current and future status of adjuvant therapy for breast cancer. *Cancer* 97: 880-886.

Coombes RC, Hall E, Gibson LJ, Paridaens R, Jassem J, Delozier T, Jones SE, Alvarez I, Bertelli G, Ortmann O, Coates AS, Bajetta E, Dodwell D, Coleman RE, Fallowfield LJ, Mickiewicz E, Andersen J, Lønning PE, Cocconi G, Stewart A, Stuart N, Snowdon CF, Carpentieri M, Massimini G, Bliss JM, van de Velde C; Intergroup Exemestane Study. (2004) A randomized trial of exemestane after two to three years of tamoxifen therapy in postmenopausal women with primary breast cancer. *N Engl J Med* 350: 1081-1092.

De Jong PC, van de Ven J, Nortier HW, Maitimu-Smeele I, Donker TH, Thijssen JH, Slee PH, Blankenstein RA. (1997) Inhibition of breast cancer tissue aromatase activity and estrogen concentrations by the third-generation aromatase inhibitor vorozole. *Cancer Res* 57: 2109-2111.

Fang ZJ, Yang S, Gurates B, Tamura M, Simpson ER, Evans D, Bulun SE. (2002) Genetic or enzymatic disruption of aromatase inhibits the growth of ectopic uterine tissue. *J Clin Endocrinol Metab* 87: 3460-3466.

Fatemi HM, Al-Turki HA, Papanikolaou EG, Kosmas L, De Sutter P, Devroey P. (2005) Successful treatment of an aggressive recurrent post-menopausal endometriosis with an aromatase inhibitor. *Reprod Biomed Online* 11: 455-457.

Forward DP, Cheung KL, Jackson L, Robertson JF. (2004) Clinical and endocrine data for goserelin plus anastrozole as second-line endocrine therapy for premenopausal advanced breast cancer. *Br J Cancer* 90: 590-594.

Giudice LC, Kao LC. (2004) Endometriosis. *Lancet* 364: 1789-1799.

Hefler L, Grimm C, van Trotsenburg M, Nagele F. (2005) Role of the vaginally administered aromatase inhibitor anastrozole in women with rectovaginal endometriosis: a pilot study. *Fertil Steril* 84: 1033-1036.

Jenkins VA, Ambroisine LM, Atkins L, Cuzzick J, Howell A, Fallowfield LJ. (2008) Effects of anastrozole on cognitive performance in postmenopausal women: a randomized, double-blind chemoprevention trial (IBIS II). *Lancet Oncol* 9: 953-961.

Kitawaki J, Noguchi T, Amatsu T, Maeda K, Tsukamoto K, Yamamoto T, Fushiki S, Osawa Y, Honjo H. (1997) Expression of aromatase cytochrome P450 protein and messenger ribonucleic acid in human endometriotic and adenomyotic tissues but not in normal endometrium. *Biol Reprod* 57: 514-519.

Gazvani R, Templeton A. (2002) New considerations for the pathogenesis of endometriosis. *Int J Gynaecol Obstet* 76: 117-126.

Goldhirsch A, Glick JH, Gelber RD, Glick JH, Thürlimann B, Senn HJ; St Gallen Expert Panel Members. (2005) Meeting highlights: International Expert Consensus on the primary therapy of early breast cancer 2005. *Ann Oncol* 161: 569-583.

Goss P, Ingle J, Martino S, Robert NJ, Muss HB, Piccart MJ, Castiglione M, Tu D, Shepherd LE, Pritchard KI, Livingston RB, Davidson NE, Norton L, Perez EA, Abrams JS, Therasse P, Palmer MJ, Pater JL. (2003) A randomized trial of letrozole in postmenopausal women after 5 years of tamoxifen therapy for early-stage breast cancer. *N Engl J Med* 349: 1-10.

Goss PE, Strasser K. (2001) Aromatase inhibitors in the treatment and prevention of breast cancer. *J Clin Oncol* 19: 881-894.

Gruber CJ, Tschugguel W, Schneeberger C, Huber JC. (2002) Production and actions of estrogens. *N Engl J Med* 346: 340-352.

Gurates B, Sebastian S, Yang S, Zhou J, Tamura M, Fang Z, Suzuki T, Sasano H and Bulun SE (2002) WT1 and DAX-1 inhibit aromatase P450 expression in human endometrial and endometriotic stromal cells. *J Clin Endocrinol Metab* 87: 4369-4377.

Henzl MR, Corson SL, Moghissi K, Buttram VC, Berqvist C, Jacobson J. (1988) Administration of nasal nafarelin as compared with oral danazol for endometriosis. *N Engl J Med* 318: 485-489.

Kao LC, Germeyer A, Tulac S, Lobo S, Yang JP, Taylor RN, Osteen KG, Lessey BA, Giudice LC. (2003) Expression profiling of endometrium from women with endometriosis reveals candidate genes for disease based implantation failure and infertility. *Endocrinology* 144: 2870-2881.

Koninckx PR, Meuleman C, Demeyere S, Lesaffre E, Cornillie FJ. (1991) Suggestive evidence that pelvic endometriosis is a progressive disease, whereas deeply infiltrating endometriosis is associated with pelvic pain. *Fertil Steril* 55: 759-765.

Lessey BA, Castelbaum AJ, Sawin SW, Buck CA, Schinnar R, Bilker W, Strom BL. (1994) Aberrant integrin expression in the endometrium of women with endometriosis. *J Clin Endocrinol Metab* 79: 643-649.

Meresman GF, Bilotas M, Abello V, Buquet R, Tesone M, Sueldo C. (2005) Effects of aromatase inhibitors on proliferation and apoptosis in eutopic endometrial cell cultures from patients with endometriosis. *Fertil Steril* 84: 459-463.

Michael MD, Kilgore MW, Morohashi KI, Simpson ER. (1995) Ad4 BP/SF-1 regulates cyclic AMP-induced transcription from the proximal promoter (PII) of the human aromatase P450 (CYP19) gene in the ovary. *J Biol Chem* 270: 13561-13566.

Michael MD, Michael LF, Simpson ER. (1997) A CRE-like sequence that binds CREB and contributes to cAMP dependent regulation of the proximal promoter of the human aromatase P450 (CYP19) gene. *Molecular and Cellular Endocrinology* 134: 147-156.

Mitwally MF, Casper RF. (2004) Aromatase inhibition reduces the dose of gonadotropin required for controlled ovarian hyperstimulation. *J Soc Gynecol Investig* 11: 406-415.

Noble LS, Simpson ER, Johns A, Bulun SE. (1996) Aromatase expression in endometriosis. *J Clin Endocrinol Metab* 81: 174-179.

Noble LS, Takayama K, Putman JM, Johns DA, Hinshelwood MM, Agarwal VR, Zhao Y, Carr BR, Bulun SE. (1997) Prostaglandin E2 stimulates aromatase expression in endometriosis-derived stromal cells. *J Clin Endocrinol Metab* 82: 600-606.

Olive DL, Lindheim SR, Pritts EA. (2004) New medical treatments for endometriosis. *Best Pract Clin Obstet Gynaecol* 18: 319-328.

Olive DL, Pritts EA. (2001) Treatment of endometriosis. *N Engl J Med* 345: 266-275.

Olive DL, Schwartz LB. (1993) Endometriosis. *N Engl J Med* 328: 1759-1769.

Patwardhan S, Nawathe A, Yates D, Harrison GR, Khan KS. (2008) Systematic review of the effects of aromatase inhibitors on pain associated with endometriosis. *Br J Obstet Gynaecol* 115: 818-822.

Plourde PV, Dyroff M, Dukes M. (1994) Arimidex: a potent and selective fourth-generation aromatase inhibitor. *Breast Cancer Res Treat* 30: 103-111.

Practice Committee of the American Society for Reproductive Medicine. (2004) Endometriosis and infertility. *Fertil Steril* 81: 1441-1446.

Prentice A, Randall BJ, Weddell A, McGill A, Henry L, Horne CHW, Thomas EJ (1992) Ovarian steroid receptor expression in endometriosis and in two potential parent epithelia: endometrium and peritoneal mesothelium. *Hum Reprod* 7: 1318-1325.

Razzi S, Fava A, Sartini A, De Simone S, Cobellis L, Petraglia F. (2004) Treatment of severe recurrent endometriosis with an aromatase inhibitor in a young ovariectomised woman. *Br J Obstet Gynaecol* 111: 182-184.

Remorgida V, Abbamonte LH, Ragni N, Fulcheri E, Ferrero S. (2007) Letrozole and desogestrel-only contraceptive pill for the treatment of stage IV endometriosis. *Aust N Z J Obstet Gynaecol* 47: 222-225.

Santen RJ. (2003) Inhibition of aromatase: insights from recent studies. Steroids 68: 559-567.

Santen RJ, Samojlik E, Wells SA. (1980) Resistance of the ovary to blockade of aromatization with aminoglutethimide. *J Clin Endocrinol Metab* 51: 473-477.

Scott LJ, Wiseman LR. (1999) Exemestane. *Drugs* 58: 675-680.

Sebastian S, Takayama K, Shouzu M, Bulun S. (2002) Cloning and characterization of a novel endothelial promoter of the human CYP19 (aromatase P450) gene that is up-regulated in breast cancer tissue. *Mol Endocrinol* 16: 2243-2254.

Sestak I, Cuzick J, Sapunar F, Eastell R, Forbes JF, Bianco AR, Buzdar AU. (2008) Risk factors for joint symptoms in patients enrolled in the ATAC trial: a retrospective, exploratory analysis. *Lancet Oncol* 9: 866-872.

Shippen ER, West WJ Jr. (2004) Successful treatment of severe endometriosis in two premenopausal women with an aromatase inhibitor. *Fertil Steril* 81: 1395-1398.

Simpson ER, Clyne C, Rubin G, Boon WC, Robertson K, Britt K, Speed C, Jones M. (2002) Aromatase – a brief overview. *Annu Rev Physiol* 64: 93-127.

Simpson ER, Mahendroo MS, Means GD, Kilgore MW, Hinshelwood MM, Graham-Lorence S, Amarney B, Ito Y, Fisher CR, Michael MD, Mendelson CR, Bulun SE. (1994) Aromatase cytochrome P450, the enzyme responsible for estrogen biosynthesis. *Endcr Rev* 15: 342-355.

Simpson ER, Zhao Y, Agarwal VR, Michael MD, Bulun SE, Hinshelwood MM, Graham-Lorence S, Sun T, Fisher CR, Qin K, Mendelson CR. (1997) Aromatase expression in health and disease. *Recent Prog Horm Res* 52: 185-213.

Soysal S, Soysal M, Ozer S, Gul N, Gezgin T. (2004) The effects of postsurgical administration of goserelin plus anastrozole compared to goserelin alone in patients with severe endometriosis: a prospective randomized trial. *Hum Reprod* 19: 160-167.

Stein RC, Dowsett M, Hedley A, Gazet JC, Ford HT, Coombes RC. (1990) The clinical and endocrine effects of 4-hydroxyandrostenedione alone and in combination with goserelin in premenopausal women with advanced breast cancer. *Br J Cancer* 62: 679-683.

Sun HS, Hsiao KY, Hsu CC, Wu MH, Tsai SJ. (2003) Transactivation of steroidogenic acute regulatory protein in human endometriotic stromal cells is mediated by the prostaglandin EP2 receptor. *Endocrinology* 144: 3934–3942.

Takayama K, Zeitoun K, Gunby RT, Sasano H, Carr BR, Bulun SE. (1998) Treatment of severe postmenopausal endometriosis with an aromatase inhibitor. *Fertil Steril* 69: 709-713.

The ATAC Trialists' Group. (2005) Results of the ATAC (Arimidex, Tamoxifen Alone or in Combination) trial after completion of 5 years' adjuvant treatment for breast cancer. *Lancet* 365: 60-62.

Tsai SJ, Wu MH, Lin CC, Sun HS, Chan HM. (2001) Regulation of steroidogenic acute regulatory protein expression and progesterone production in endometriotic stromal cells. *J Clin Endocrinol Metab* 86: 5765-5773.

Vercellini P, Cortesi I, Crosignani PG (1997) Progestins for symptomatic endometriosis: a critical analysis of the evidence. *Fertil Steril* 68: 393-401.

Vessey MP, Villard-Mackintosh L, Painter R. (1993) Epidemiology of endometriosis in women attending family planning clinics. *Br Med J* 306: 182-184.

Viganò P, Mangioni S, Odorizzi MP, Chiodini A, Rocca S, Chiodo I (2003) Use of estrogen antagonists and aromatase inhibitors in endometriosis. *Curr Opin Investig Drugs 4:* 1209-1212.

Waller KG, Shaw RW. (1993) Gonadotropin-releasing hormone analogues for the treatment of endometriosis: long-term follow-up. *Fertil Steril* 59: 511-515.

Wheeler JM. (1989) Epidemiology of endometriosis associated infertility. *J Reprod Med* 34: 41-46.

Winer EP, Hudis C, Burstein HJ, Chlebowski RT, Ingle JN, Edge SB, Mamounas EP, Gralow J, Goldstein LJ, Pritchard KI, Braun S, Cobleigh MA, Langer AS, Perotti J, Powles TJ, Whelan TJ, Browman GP. (2002) American Society of Clinical Oncology technology assessment on the use of aromatase inhibitors as adjuvant therapy for women with hormone receptor-positive breast cancer: status report 2002. *J Clin Oncol* 20: 3317-3327.

Wu M, Wang C, Lin C, Chen L, Chang W, Tsai S. (2005) Distinct regulation of cyclooxygenase-2 by interleukin-1β in normal and endometriotic stromal cells. *J Clin Endocrinol Metab* 90: 286-295.

Yang S, Fang Z, Takashi S, Hironobu S, Jianfeng Z, Gurates B, Tamura M, Ferrer K, Bulun SE. (2002) Regulation of aromatase P450 expression in endometriotic and endometrial stromal cells by CCAT/enhancer binding proteins: decreased C/EBPβ in endometriosis is associated with overexpression of aromatase. *J Clin Endocrinol Metab* 87: 2336-2345.

Zeitoun KM, Bulun SE. (1999) Aromatase: a key molecule in the pathophysiology of endometriosis and a therapeutic target. *Fertil Steril* 72: 961-969.

Zeitoun K, Takayama K, Michael MD, Bulun SE. (1999) Stimulation of aromatase P450 promoter (II) activity in endometriosis and its inhibition in endometrium are regulated by competitive binding of steroidogenic factor-1 and chicken ovalbumin upstream promoter transcription factor to the same cis-acting element. *Mol Endocrinol* 13: 239-253

Zeitoun K, Takayama K, Sasano H, Suzuki T, Moghrabi N, Andersson S, Johns A, Meng L, Putman M, Carr B, Bulun SE. (1998) Deficient 17beta-hydroxysteroid dehydrogenase type 2 expression in endometriosis: failure to metabolize 17beta-estradiol. *J Clin Endocrinol Metab* 83: 4474-4480.

In: Aromatase Inhibitors: Types, Mode of Action and Indications ISBN: 978-1-60741-711-8
Editor: Jean R. Lamonte

Chapter VII

An Update of the Use of Aromatase Inhibitors in the Management of Breast Cancer

S. Tang, D. A. L. Morgan[1] and K. L. Cheung*
Division of Breast Surgery, University of Nottingham;
[1]Department of Oncology, Nottingham University Hospitals, UK

Abstract

With the increasing use of endocrine therapy in the treatment of breast cancer, the role of aromatase inhibitors in the management of hormone receptor positive postmenopausal breast cancer has been thoroughly investigated. Third generation aromatase inhibitors (anastrozole, letrozole and exemestane) are very well tolerated, with potentially manageable side effects. They have an established role in the treatment of advanced disease. They have also been shown to be more effective than tamoxifen when used as adjuvant treatment in early primary breast cancer, either when used upfront, as sequential / switch therapy after two to three years of tamoxifen or as extended therapy after five years of tamoxifen. Recent studies of their use in premenopausal women (alongside ovarian suppression therapy) and as neoadjuvant treatment have shown promising results. Their potential roles in chemoprevention and as combination therapies (with biological therapy or a pure anti-oestrogen) are being investigated in ongoing trials. Further studies need to be conducted to identify tumour and patient prognostic and predictive factors that may improve patient selection for optimal targeted therapy with aromatase inhibitors.

* *Corresponding author:* Mr KL Cheung. Clinical Associate Professor / Consultant Breast Surgeon, Professorial Unit of Surgery, City Hospital, Hucknall Road, Nottingham NG5 1PB. UK. Tel: +44 (0) 115 8231876; Fax: +44 (0) 115 8231877; Email: kl.cheung@nottingham.ac.uk

Introduction

The role of endocrine therapy for breast cancer was discovered more than a century ago, when Sir George Beatson reported performing the first oophorectomy in a patient with inoperable breast cancer leading to tumour regression[1]. Since then, research has progressed to discover the oestrogen receptors and their functions and role in breast cancer. The presence of oestrogen receptor lends itself to the development of an alternative treatment modality in breast cancer, besides surgery, radiotherapy and chemotherapy.

Endocrine therapy was initially carried out through the removal of sources of oestrogen, either by surgery (eg.oophorectomy, adrenalectomy or even pituitary ablation) or by radiotherapy (ie. to the ovaries). Currently, it is commonly achieved through manipulation by pharmacological agents (eg. tamoxifen, aromatase inhibitors, gonadotrophin or luteinizing hormone releasing hormone (GnRH or LHRH) agonist) which have the added benefit for younger patients of being reversible. Perimenopausal patients often prefer a permanent treatment. These pharmacological agents work via the antagonism of oestrogen or the manipulation of oestrogen receptors.

Oestrogens produce normal physiological effects by binding to nuclear receptor proteins. At present two oestrogen receptors, α and β, are known. The predominant oestrogen receptor in the female reproductive tract and mammary glands is oestrogen receptor-α[2]. It is a 65kDA oestrogen-binding protein expressed by 46-77% of breast cancers[3-7] and is more likely to be expressed in breast cancers in elderly patients[7-9]. The oestrogen receptor remains the strongest predictor of response to endocrine therapy. Pharmacological agents in endocrine therapy can act directly on the oestrogen receptor of an oestrogen receptor positive tumour, or deprive the tumour of circulating oestrogen (Figure 1). From historical studies, up to 60% of oestrogen receptor positive tumours and 10% of oestrogen receptor negative tumours respond to endocrine therapy[10-13]. However, the degree of oestrogen receptor positivity is often not clearly defined in these studies. Furthermore, it is thought that with the development of better assays for the detection of oestrogen receptors, previously presumed oestrogen receptor negative tumours may indeed be weakly oestrogen receptor positive. As such, the rate of response of oestrogen receptor positive tumours to endocrine therapy may indeed be higher, with a lower response rate in true oestrogen receptor negative tumours.

The discovery of tamoxifen in 1971 has revolutionised the use of endocrine therapy for breast cancer[14]. It is a selective oestrogen receptor modulator(SERM) which has both oestrogen agonistic and antagonistic properties. The anti-oestrogenic effect competitively inhibits the oestrogen receptors in breast cancer, which in turn inhibits the proliferation of breast cancer cells. Until recently, it has been the gold standard for endocrine therapy in breast cancer and was the standard against which newer agents were compared. It is used in advanced breast cancer, adjuvant and neoadjuvant phases of treatment for oestrogen receptor positive breast cancers in both pre- and postmenopausal women, and has been licensed in the United States for chemoprevention in high risk women.

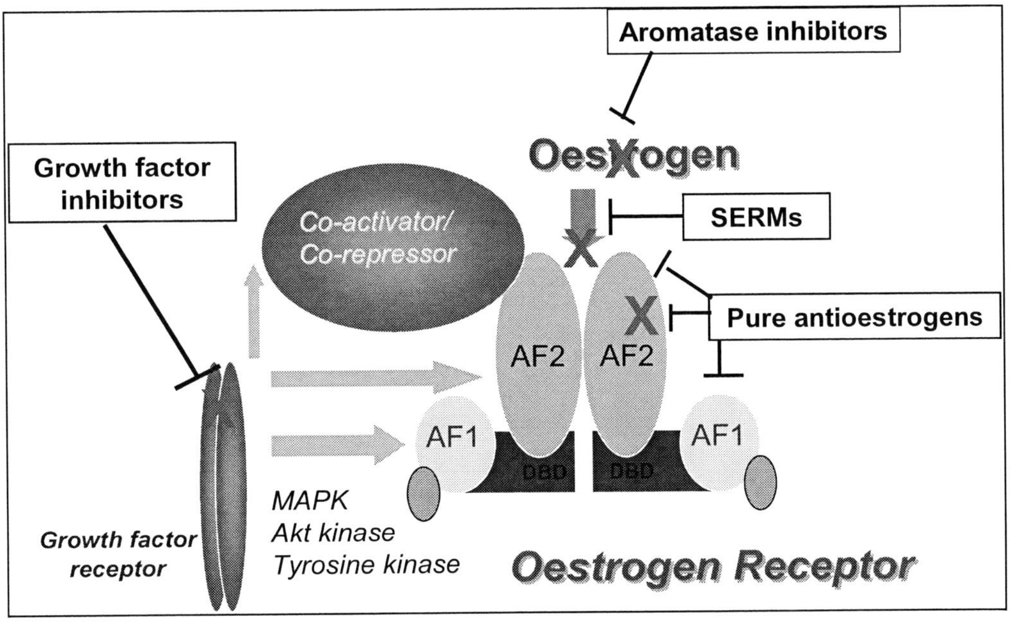

Abbreviations:
SERMs = selective oestrogen receptor modulators
AF = Activation functions
DBD = DNA-binding domain
MAPK = Mitogen-activated protein kinase

Figure 1. Targeting the oestrogen receptor pathway.

Aromatase Inhibitors

The development of third generation aromatase inhibitors has challenged the role of tamoxifen as the gold standard for endocrine therapy in breast cancer. It acts by blocking the production of oestrogen via the aromatase system. The aromatase enzyme is part of the cytochrome P450 family and plays a crucial role in the synthesis of oestrogen. It controls the conversion of testosterone to oestradiol (E_2) directly and also the conversion of androstenedione to oestrone (Figure 2). In premenopausal women, most aromatase activity is in the granulosa cells of the ovary, where most oestrogens are produced. After menopause, ovarian granulosa cells cease to function, and the main source of oestrogen (and aromatase activity) is in non-ovarian tissue eg. fat, adrenals, liver.

Studies have demonstrated that nearly two-thirds of breast cancer cells express aromatase with *in vitro* activity[15-18]. The aromatase levels in breast tumours are higher than in the peritumoural fat. Furthermore, the aromatase levels in the quadrant where the tumour is located, is higher than in other quadrants within the breast. This leads to an oestrogen-rich microenvironment around the tumour. As such, aromatase inhibitors act both locally and systemically to deprive breast tumours of oestrogen, inhibiting breast cancer cell proliferation.

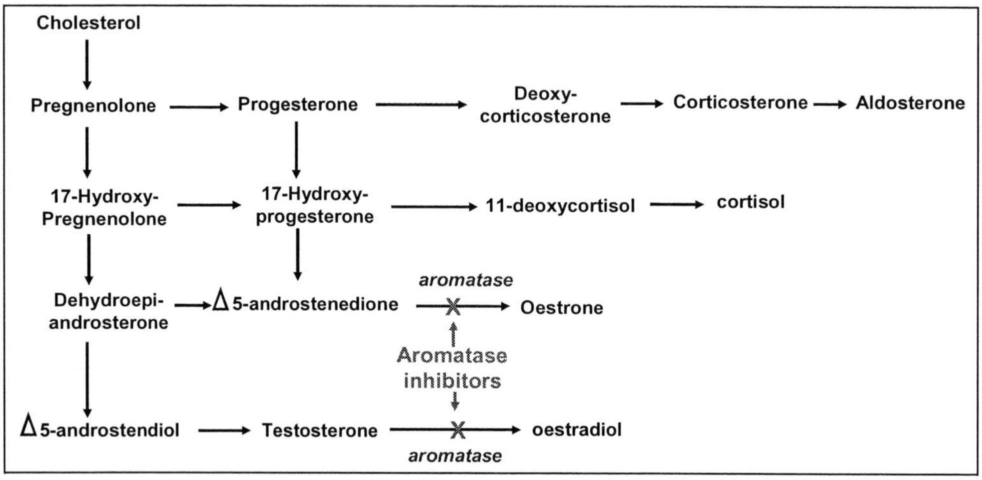

Figure 2. Oestrogen synthesis pathway, showing level of action of aromatase inhibitors.

There are three generations of aromatase inhibitors based on the timing of their development. Aminoglutethimide is the only first generation non-steroidal aromatase inhibitor[19]. It is non-selective and inhibits both glucocorticoid and mineralocorticoid synthesis; thus patients on aminoglutethimide require steroid supplementation. It was originally used as an antiepileptic agent but was removed from the market due to serious side effects. However, it became the first aromatase inhibitor evaluated in clinical studies for the treatment of hormone-dependent breast cancer in the 1970s[20, 21]. Numerous clinical trials have demonstrated the clinical efficacy of aminoglutethimide (with corticosteroid replacement) in the treatment of breast cancer. However, the side effects of lethargy and morbilliform skin rash (Figure 3), with the development of newer more potent aromatase inhibitors, resulted in the cessation of the use of aminoglutethimide in breast cancer.

Second-generation aromatase inhibitors include formestane and fadrozole. They are more selective than aminoglutethimide and its inhibitory activity is 700 times more potent[2]. Although they had some non-selective inhibitory activity, they were well tolerated with a better side effect profile than aminoglutethimide. Clinical trials showed them to be effective in reducing plasma oestrogen levels and in causing partial or complete regression of tumours in patients with advanced breast cancer[22]. However, it is administered through the intramuscular route, which limited its ease of use.

The development of third generation selective aromatase inhibitors rapidly superceded the use of formestane and fadrozole in breast cancer; as they are more potent, available orally, with even better side effect profiles. There are type I steroidal (exemestane) and type II non-steroidal (anastrozole and letrozole) third generation aromatase inhibitors (Figure 4). All three are taken daily with good oral bioavailability. They all produce >95% inhibition of oestrogen biosynthesis in patients[23-25]. Letrozole has been shown to produce the greatest suppression (up to 99%), but this has not been translated into significantly greater clinical response compared to anastrozole and exemestane[26, 27]. The main side effects reported are loss in bone mineral density and altered lipid profile, which are potentially manageable. Currently, third generation aromatase inhibitors are indicated in advanced breast cancer, adjuvant and neoadjuvant treatment of hormone receptor positive breast cancer in postmenopausal patients.

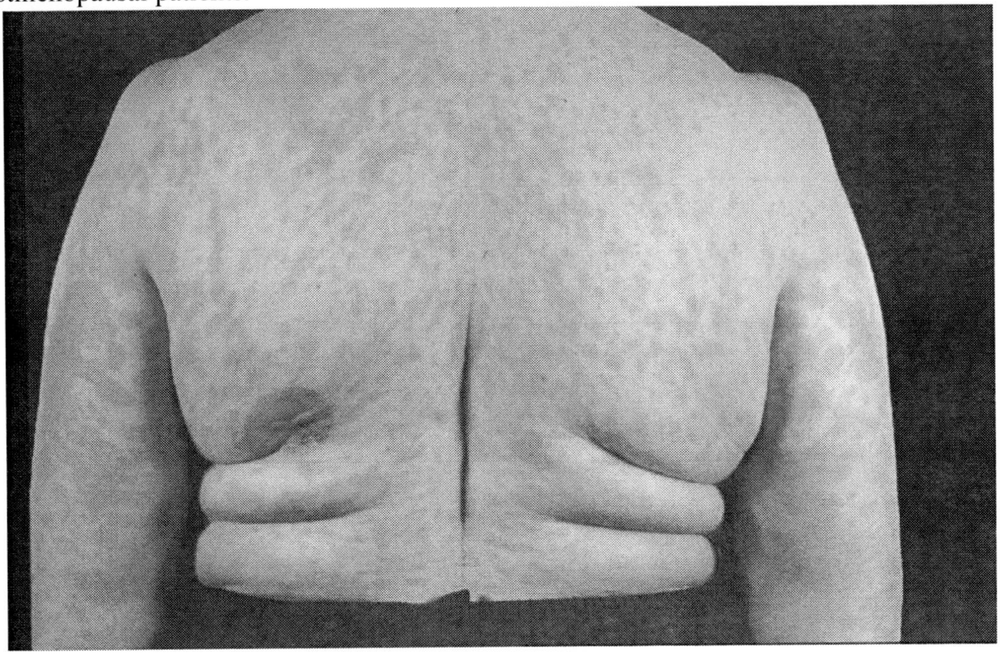

Figure 3. Morbilliform skin rash – side effect of aminoglutethimide.

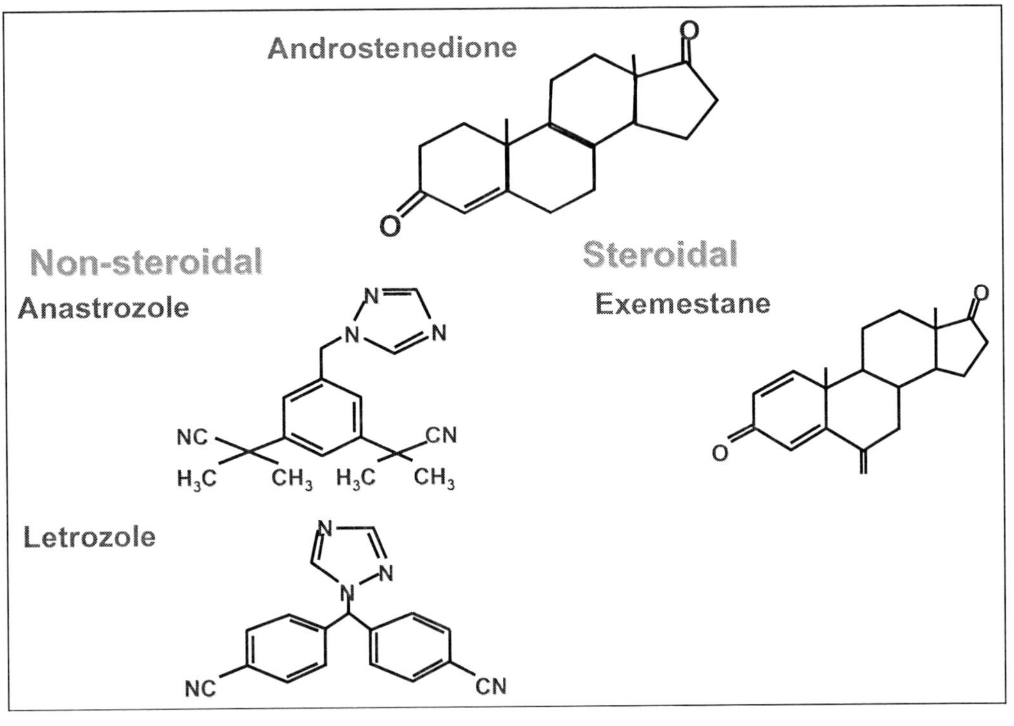

Figure 4. The molecular structure of androstenedione compared with type I (steroidal) and type II (non-steroidal) third generation aromatase inhibitors.

Aromatase Inhibitors in Advanced Cancer

Third generation aromatase inhibitors were first introduced for the use in advanced breast cancer. They were compared with megestrol acetate as second line therapy in advanced breast cancer in women who failed to respond to tamoxifen. Results from numerous clinical trials showed all third generation aromatase inhibitors to be as effective as, if not superior to megestrol acetate. They were found to have improved objective response rates, produced longer duration of clinical benefit with some improvement in overall survival[28-31] (Table 1). They were also better tolerated by patients with an improved side effect profile.

Table 1. Megestrol acetate vs aromatase inhibitors in advanced breast cancer in patients who failed to respond to tamoxifen

	Anastrozole	Megestrol acetate	Letrozole	Megestrol acetate	Exemestane	Megestrol acetate
	1mg,od	40mg,qid	2.5mg,od	40mg,qid	25mg,od	40mg,qid
	Europe & US studies		European study only		International study	
	n=263	n=253	n=174	n=189	n=366	n=403
OR (CR+PR)	12.5%	12.2%	23.6%*	16.4%	15.0%	12.4%
Clinical benefit (CR+PR+SD>24 wk)	42.2%	40.3%	34.5%	31.7%	37.4%	34.6%
Duration of clinical benefit (mos)	18.3	15.7	24*	14	13.8*	11.3
Follow up	31 months		33 months		11 months	
Endpoint						
TTP (mos)	4.8	4.6	5.6	5.5	4.7 (p=0.037)	3.8
Survival (mos)	26.7 (p<0.025)	22.5	25.3	21.5	NR (p=0.039)	28.4

* Statistically significant.
Abbreviations:
OR = objective response
SD = stable disease
mos = months
CR = complete response
TTP = time to progression
od = once daily
PR = partial response
NR = not reached
qid = four times daily

Table 2. Outcome from trials comparing anastrozole versus tamoxifen as first line systemic endocrine therapy in advanced breast cancer

	TARGET trial		North American trial		Combined analysis	
	Anastrozole 1mg (n=340)	Tamoxifen 20mg (n=328)	Anastrozole 1mg (n=171)	Tamoxifen 20mg (n=182)	Anastrozole 1mg (n=511)	Tamoxifen 20mg (n=510)
ER / PR positive(%)	45.3	43.9	88.3	89	59.7	60
OR (CR + PR) (%)	32.9	32.6	21.1	17	29	27
CB (CR + PR + SD> 24 weeks) (%)	56.1	55.5	59.1	45.6	57	52
Median TTP in all patients (mos)	8.2	8.3	11.1*	5.6	8.5	7.0
Median TTP in ER / PR positive patients (mos)	8.9	7.8	Not reported	Not reported	10.7*	6.4

* Statistically significant.

Abbreviations:

TARGET = Tamoxifen or Arimidex Randomised Group Efficacy and Tolerability
ER = oestrogen receptor
CB = clinical benefit
SD = stable disease
PR = progesterone receptor
CR = complete response
TTP = time to progression
OR = objective response
PR = partial response
mos = months

Following this, two large double-blind randomised controlled trials were designed to compare third generation aromatase inhibitors with tamoxifen (the standard first line treatment for advanced breast cancer at that time). This was carried out separately in North America[32], and in the rest of the world (Tamoxifen or Arimidex Randomised Group Efficacy and Tolerability (TARGET) trial)[33]. Patients were randomised to receive either anastrozole 1mg daily (with placebo tamoxifen) or tamoxifen 20mg daily (with placebo anastrozole). The primary end points were time to progression, objective response and tolerability; and the secondary end points were time to treatment failure, time to progression and survival. Nabholtz et al reported that the median time to progression was increased to 11.1 months by anastrozole, as opposed to 5.6 months with tamoxifen. However, this finding was not replicated in the TARGET trial whereby anastrozole and tamoxifen had similar efficacy and had equivalent time to progression - 8.2 months (anastrozole) vs 8.3 months (tamoxifen)[33]. This discrepancy may be explained by the difference in hormone receptor status of the patients in the trials, with the North American trial having nearly 90% of oestrogen receptor positive patients, whilst in the TARGET trial <50% of patients were oestrogen receptor positive (Table 2). In the combined analysis of both studies, patients in the oestrogen receptor positive subgroup had a significantly better response to anastrozole, with median time to progression of 10.7 months, versus 6.4 months with tamoxifen.

Further subgroup analysis was carried out on patients who progressed on their primary treatment and crossed over from anastrozole to tamoxifen or vice versa. The clinical outcomes for both groups of patients were similar, suggesting that tamoxifen is equally effective as second line therapy in patients who have progressed on anastrozole; whereby anastrozole does not induce resistance to tamoxifen[34].

Both these trials compared the tolerability of anastrozole and tamoxifen among the patient population, with predefined adverse events identified based on the known pharmacology and expected side effects of both treatments. The main side effects recorded with aromatase inhibitors are related to the withdrawal of oestrogen – ie. hot flushes and vaginal dryness. However, they have a safer side effect profile compared to tamoxifen, with anastrozole being associated with fewer incidences of thromboembolic events and vaginal bleeding [33] at median follow-up of 18 months (Table 3).

In a separate study, letrozole was compared with tamoxifen as first line treatment in advanced breast cancer. Patients were randomised to receive either letrozole 2.5mg or tamoxifen 20mg daily. Treatment was continued until disease progression, or discontinuation for other reasons; whereby patients may then crossover to the alternative treatment (if still suitable for endocrine treatment). Patients were followed up for disease progression and survival. Patients treated with letrozole had a significantly improved clinical benefit of 49% compared with tamoxifen (38%) (p=0.001). These patients also demonstrated improved time to progression on letrozole (9.4 vs 6.0 months; p<0.0001) (Table 4)[35].

Table 3. Tolerability data from combined analysis of the North American and Tamoxifen or Arimidex Randomised Group Efficacy and Tolerability (TARGET) trials at median follow-up of 18.2 months for predefined adverse events

	Anastrozole 1mg (*n*=506)		Tamoxifen 20mg (*n*=511)		Critical Significance Level
	n	%	*n*	%	
Weight gain	11	2.2	8	1.6	0.0167
Gastrointestinal disturbance	170	33.6	196	38.4	0.0071
Hot flushes	134	26.5	118	23.1	0.0100
Tumour flare	15	3.0	18	3.5	0.0500
Thromboembolic disease	23	4.5	39	7.6	0.0056
Lethargy	6	1.2	15	2.9	0.0063
Depression	23	4.5	32	6.3	0.0125
Vaginal bleeding	5	1.0	11	2.2	0.0083
Vaginal dryness	15	3.0	13	2.5	0.0250

Table 4. Letrozole versus tamoxifen as first line endocrine therapy for advanced breast cancer

	Letrozole 2.5mg (N= 453)	Tamoxifen 20mg (N= 454)
Median TTP (mos)	9.4*	6.0
Median TTF (mos)	9.1*	5.7
OR (CR + PR) (%)	30*	20
Clinical Benefit (CR+PR+SD>24 wk) (%)	49*	38

* Statistically significant.
Abbreviations:
TTP = time to tumour progression
OR = objective response
mos = months
TTF = time to treatment failure
CR = complete response
PR = partial response
SD = stable disease

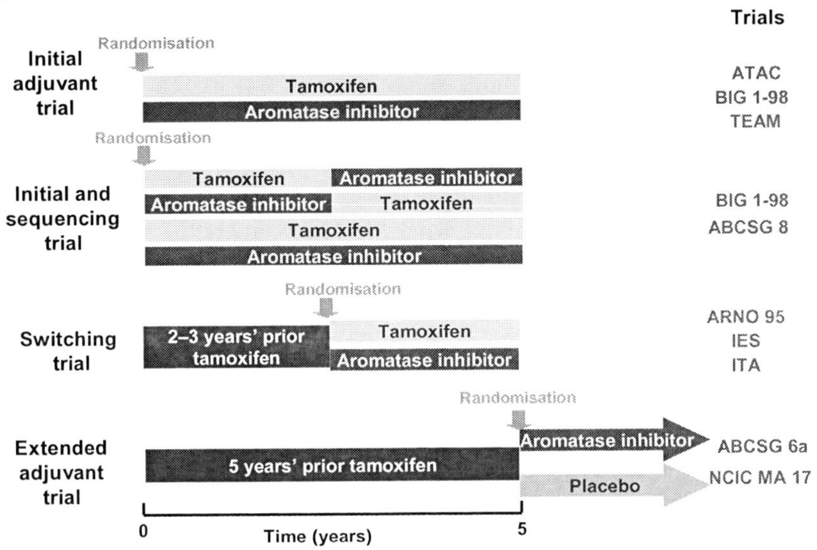

Abbreviations:
ATAC = Anastrozole, Tamoxifen, Alone or in Combination trial
BIG 1-98 = Breast International Group 1-98 trial
TEAM = Tamoxifen, Exemestane Adjuvant Multinational trial
ABCSG 8 = Austrian Breast and Colorectal cancer Study Group trial 8
ARNO 95 = Arimidex-Nolvadex 95 trial
IES = International Exemestane Study
ITA = Italian Tamoxifen Anastrazole trial
ABCSG 6a = Austrian Breast and Colorectal cancer Study Group trial 6a
NCIC MA 17 = National Cancer Institute of Canada clinical trials group trial MA 17

Figure 5. Study design of different randomised trials investigating the use of aromatase inhibitors in adjuvant endocrine therapy in postmenopausal women.

Similarly, a recently concluded multicentre randomised controlled trial compared exemestane 25mg with tamoxifen 20mg as first line endocrine treatment in post-menopausal women with metastatic breast cancer. It demonstrated an improved response rate in exemestane treated patients (46% vs 31%; p=0.005) with an increased median time to progression of 9.9 months (vs 5.8 months with Tamoxifen)[36].

The results of the various trials support the use of third generation aromatase inhibitors in advanced breast cancer. They have been shown to be better than megestrol acetate as second line therapy, and are better than tamoxifen as first line therapy in postmenopausal women with oestrogen receptor positive advanced breast cancer. This is true for both efficacy and tolerability in these patients. As such, it should be considered for first line endocrine therapy in postmenopausal women with oestrogen receptor positive advanced breast cancer.

Aromatase Inhibitors as Adjuvant Therapy

With the success of third generation aromatase inhibitors in advanced breast cancer, research has expanded to investigate their potential role in the adjuvant setting. Tamoxifen has previously been established as the gold standard in endocrine therapy, with a key role in adjuvant therapy. However, its use has been limited by life threatening side effects of endometrial cancer, thromboembolic events (pulmonary embolism) and cerebrovascular events (stroke). As such, aromatase inhibitors may fulfill a crucial role in adjuvant therapy for patients with increased risk of these side effects, as they have a more favourable side effect profile and have been well tolerated during their use in advanced breast cancer. The trials designed to evaluate the use of aromatase inhibitors in the adjuvant setting have taken two main approaches – the aromatase inhibitors are either compared with tamoxifen directly, or are used sequentially with Tamoxifen. The different study designs and randomisation points are summarised in Figure 5.

Initial Adjuvant Therapy

Breast cancer recurrence peaks in the first two years following surgery, with another smaller peak more than five years later (Figure 6)[37]. Trials with initial adjuvant therapy aim to target this time period where patients are at greatest risk of recurrence, as finding the best possible therapy at this stage is paramount to improving disease-free survival. The Anastrozole, Tamoxifen, Alone or in Combination (ATAC) trial is the largest randomised adjuvant trial thus far, having enrolled 9366 postmenopausal women. They were randomised into three arms, receiving either anastrozole 1mg daily, tamoxifen 20mg daily or both anastrozole and tamoxifen for five years. The combination arm was discontinued after early analysis when it showed no efficacy or tolerability benefits over tamoxifen alone. After a median follow up of 100 months, anastrozole was consistently superior to tamoxifen in nearly all the efficacy end-points (Figure 7). This was more marked in patients who had oestrogen receptor positive tumours, where they have a significantly greater disease-free survival with anastrozole compared to tamoxifen, with an absolute benefit of 2.8% between treatment arms at five years, increasing to 4% at nine years[38, 39]. This risk reduction seen in anastrozole is in addition to the benefits already seen with tamoxifen, as reported by the Early Breast Cancer Trialists' Collaborative Group (EBCTCG), where patients without adjuvant treatment had a 38% risk of recurrence, which can be reduced by 50% with tamoxifen[40]. Though there was no significant difference in overall survival between anastrozole and tamoxifen treatment groups at the end of the trial (HR = 1.00, p=0.99), the improvement in time to recurrence and distant recurrence with anastrozole suggests that an improvement in overall survival may emerge with a longer follow-up.

In the ATAC study, anastrozole was associated with fewer adverse events leading to significantly fewer women withdrawing from the trial. There were significantly fewer incidences of endometrial cancer, thromboembolic events, hot flushes, vaginal bleeding and discharge and ischemic cerebrovascular events in the anastrozole group compared with tamoxifen (Table 5)[38]. However, there were more joint problems and fractures with anastrozole. This could be accounted for by its oestrogen deprivation effects, in contrast with tamoxifen, which has oestrogen agonist property on the bones and endometrium. Of note, however, is that 7.7% of patients on tamoxifen still reported one or more fractures at different sites, with no significant difference in the incidence of hip fractures (major marker of morbidity and mortality in fractures) between both arms (Table 5)[38]. Analysis of long term follow up data showed that although fracture rates were increased whilst treated with anastrozole, no excess was noted after the five year treatment[39]. Further analysis of bone mineral density in the ATAC trial population suggests that adjuvant anastrozole therapy in postmenopausal women leads to accelerated bone loss. However, the risk of developing treatment related osteoporosis seems to be limited to women who were already osteopenic prior to treatment[41]. This could be effectively managed with intravenous or oral bisphosphonates to maintain bone mineral density whilst on endocrine therapy[42, 43].

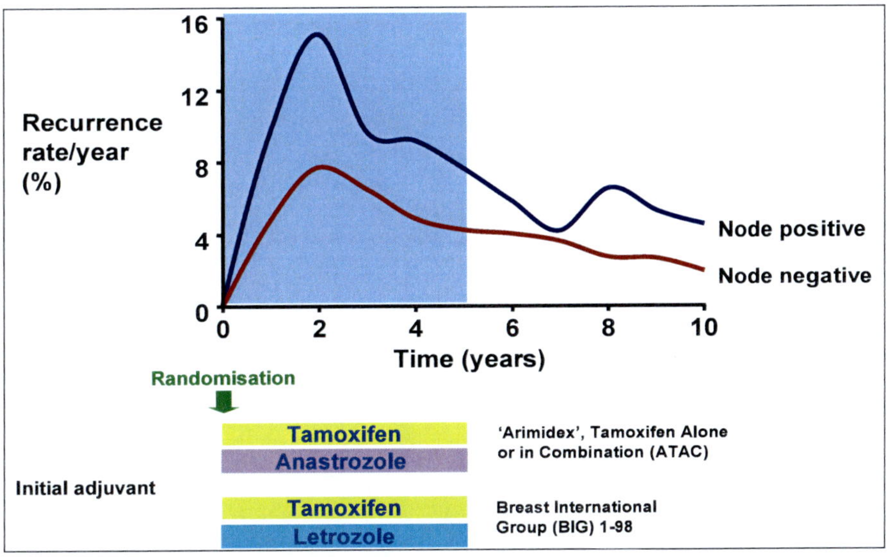

Figure 6. Recurrence pattern of breast cancer after primary treatment, with study design of initial adjuvant trials.

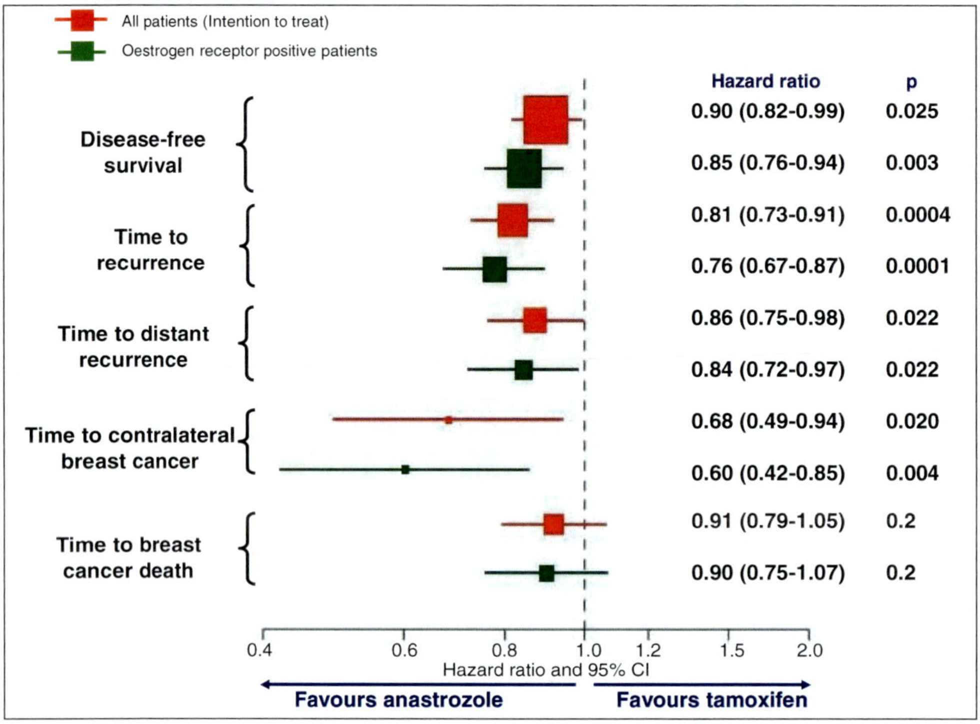

Figure 7. Efficacy analysis of the Anastrozole, Tamoxifen, Alone or in Combination (ATAC) trial at 100 months, comparing anastrozole vs tamoxifen as initial adjuvant therapy of breast cancer in postmenopausal women.

Table 5. Frequency of pre-specified adverse events in the Anastrozole, Tamoxifen, Alone or in Combination (ATAC) trial at 100 months

	Anastrozole (%)	Tamoxifen (%)
Hot flushes	36	41
Vaginal bleeding	5	10
Vaginal discharge	4	13
Endometrial cancer	0.2	0.8
Ischaemic cerebrovascular	2	3
Venous thromboembolic	3	5
Joint symptoms	36	30
Fractures	11.0	7.7
Hip	1.6	1.4
Spine	1.5	0.9
Wrist / Colles	2.3	2.0
Others	7.1	4.6
Hysterectomy	1.3	5.1

The Breast International Group (BIG) 1-98 trial was a large randomised controlled trial which compared letrozole with tamoxifen, as monotherapy or with sequential use, as adjuvant therapy in 8028 postmenopausal women with oestrogen receptor positive early breast cancer; 4922 women were assigned to adjuvant treatment with either letrozole or tamoxifen as monotherapy for five years. Analysis of this group of patients at 51 months follow up showed letrozole to be superior to tamoxifen in all the efficacy endpoints, with significant improvement in disease free survival and time to distant recurrence(Figure 8)[44].

A recent update of the trial results was presented at the San Antonio Breast Cancer Symposium in 2008. It reported the unblinding of the tamoxifen arm after earlier analysis confirmed the superiority of letrozole over tamoxifen. This led to 25.2% of patients in the tamoxifen arm crossing over to letrozole (median duration of treatment with letrozole after crossover = 18 months). Analysis of the intention to treat population showed the clinical benefit of letrozole over tamoxifen persisted at 76 months, with significant improvement in disease free survival (HR=0.88, p=0.03) and time to distant recurrence (HR=0.85, p=0.05). However, there was no significant difference in overall survival in both groups of patients[45].

The side effect profile of letrozole is similar to that of anastrozole, with similar increase in musculoskeletal side effects and fractures; with lower incidences of hot flushes, vaginal bleeding and venous thromboembolic events (Table 6). However, of concern is the increased incidence of hypercholesterolaemia in patients treated with letrozole and a trend of increased cardiac events (134 vs 122) in patients treated with letrozole; which was not reported with anastrozole[44].

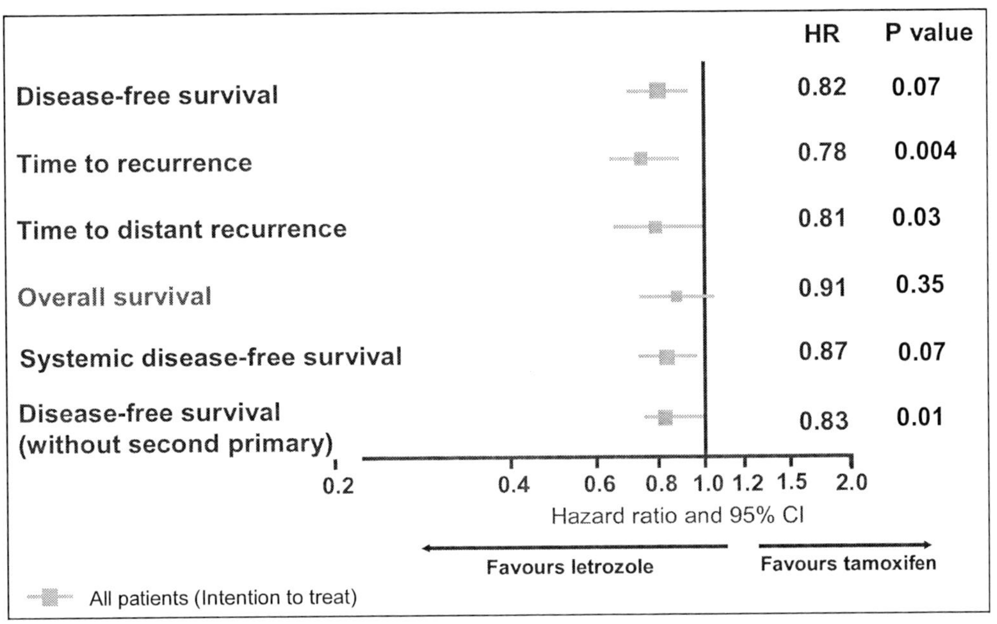

Figure 8. Efficacy analysis of primary and secondary endpoints of the Breast International Group (BIG) 1-98 trial at median follow up time of 51 months.

Table 6. Frequency of adverse events in the Breast International Group (BIG) 1-98 trial at median follow up of 51 months

	Letrozole 2.5mg (*n*=2448)		Tamoxifen 20mg (*n*=2447)		Critical Significance Level (P)
	n	%	*n*	%	
CVA / TIA	34	1.4	35	1.4	0.90
Cardiac event	134	5.5	122	5.0	0.48
Hypercholesterolaemia	1238	50.6	601	24.6	<0.001
Thromboembolic event	50	2.0	94	3.8	<0.001
Bone fractures	211	8.6	141	5.8	<0.001
Arthralgia	489	20.0	331	13.5	<0.001
Hot flushes	803	32.8	914	37.4	<0.001
Vaginal bleeding	92	3.8	203	8.3	<0.001

Abbreviations:
CVA = cerebrovascular accident
TIA = transient ischaemic attack

The Tamoxifen Exemestane Adjuvant Multinational (TEAM) trial recently presented their first planned analysis results at the San Antonio Breast Cancer Symposium. Patients were initially randomised to receive either 5 years of tamoxifen 20mg or exemestane 25mg daily. However, the protocol was amended to sequential therapy of two and a half years of tamoxifen followed by two and a half years of exemestane, as the International Exemestane Study (to be discussed later) has proven the benefit of switching from tamoxifen to exemestane after two and a half years (over 5 years of tamoxifen). All patients included in the study were postmenopausal women with oestrogen receptor or progesterone receptor positive invasive breast cancer. Analysis at 2.75 years (prior to sequencing of tamoxifen to exemestane) showed no significant difference in disease-free survival in both arms (HR=0.89, p=0.12). However, patients treated in the exemestane arm had improved relapse free survival (HR=0.85, p=0.02) and improved time to distant metastases (HR=0.81, p=<0.02). Similar to the other aromatase inhibitors, exemestane is associated with lower gynaecological side effects (eg. vaginal discharge, vaginal bleeding, endometrial hyperplasia) and thromboembolic disease; but had higher musculoskeletal side effects of arthralgia, arthritis and osteoporosis. There was no significant difference between treatment arms for cardiac events and bone fractures[46]. The final analysis of the results of the trial comparing five years of exemestane alone with tamoxifen-exemestane switching (at 2.5 years) is awaited.

These large multinational phase III randomised controlled trials have proven the efficacy of third generation aromatase inhibitors over tamoxifen as first line adjuvant monotherapy in oestrogen receptor positive early breast cancer in postmenopausal women. They are also better tolerated, with a manageable side effect profile compared to tamoxifen.

Sequential Adjuvant Therapy

Randomised trials have investigated the sequential use of aromatase inhibitors, by prospectively planning for initial treatment with tamoxifen (two to three years) followed by aromatase inhibitors (anastrozole) for a further two to three years, where patients are randomised prior to starting any adjuvant endocrine therapy (Figure 5). The Austrian Breast and Colorectal Cancer Study Group Trial 8 (ABCSG-8) showed that planned sequencing with anastrozole after two to three years of tamoxifen is beneficial, with 21% improvement in recurrence free survival (HR=0.79, p=0.038). Sequential endocrine treatment also significantly improved overall survival, with 23% decrease in death rates (HR=0.77, p=0.025)[47]. Higher levels of oestrogen receptor or progesterone receptor were predictive of longer duration of recurrence free survival and overall survival. Planned sequencing after two to three years of tamoxifen to anastrozole is superior to five years of tamoxifen treatment.

The results from the planned sequencing arms of the BIG 1-98 trial has also recently been presented. Patients were blinded into three different treatment groups – tamoxifen → letrozole, letrozole → tamoxifen (change after two to three years of initial therapy), and letrozole alone. At 71 months follow up, there was no difference in disease free survival in the sequential treatment groups compared to letrozole monotherapy. Trends suggest that patients in the letrozole → tamoxifen group had improved time to distant recurrence compared to the other treatment groups. This is more marked in patients with node positive disease. These results suggest that patients at higher risk of early recurrence should initially be treated with letrozole (over tamoxifen) as adjuvant endocrine therapy. After two years to of letrozole, they can be switched to tamoxifen (if required)[45].

Recent data has shown that sequencing of endocrine therapy is superior to tamoxifen monotherapy, but studies comparing it with aromatase inhibitor monotherapy have been inconclusive. Trends have suggested that patients (with node positive disease) at increased risk of early recurrence should be considered for an aromatase inhibitor as initial adjuvant endocrine therapy, with potential sequencing to tamoxifen after two to three years.

'Switch' Adjuvant Therapy

'Switch' adjuvant therapy clinical trials are different from sequencing trials, in that the patients have different randomisation points (Figure 5). Patients were recruited after having completed two to three years of adjuvant tamoxifen (and remained disease free), and randomised into the different treatment arms of these trials. Three switching trials have evaluated the benefit of switching patients after two to three years of tamoxifen treatment to an aromatase inhibitor compared with maintaining tamoxifen therapy for the full five years. Two trials, the Italian Tamoxifen Anastrozole (ITA) study and the Austrian Breast & Colorectal Cancer Study group (ABCSG) 8/Arimidex-Nolvadex (ARNO) 95 study switched patients to anastrozole whilst the International Exemestane Study (IES) switched patients to exemestane.

The ITA trial in 2003 was the first study to evaluate whether switching adjuvant treatment to anastrozole would benefit women with node positive early breast cancer who had already received two to three years of tamoxifen. In this study, the group allocated to switch therapy to anastrozole experienced significantly lower incidences of first events (HR=0.57; p=0.005) and recurrences (HR=0.56; p=0.01) compared with the group continuing on tamoxifen, including patients receiving adjuvant chemotherapy[48]. Analysis of the ABCSG 8 results only included data in the 3 years after the switch to anastrozole (as opposed to sequencing study above). When combined with the ARNO 95 trial, results favoured switching to anastrozole after two to three years of tamoxifen. Patients switching to anastrozole had significantly better event free survival (HR=0.6, p=0.0009), with 39% decrease in risk of metastases[49]. A meta-analysis of the results of all three trials showed that patients switching from tamoxifen to anastrozole had fewer disease recurrences, which translated to significant improvements in time to recurrence, time to distant recurrence, event-free survival and overall survival (Figure 9)[50].

The International Exemestane Study (IES) recently reported the results of the largest study to date investigating switching from tamoxifen to an aromatase inhibitor compared with completing adjuvant therapy with tamoxifen. Patients were randomised to switch to exemestane after completing two to three years of tamoxifen. After a median follow up of 55.7 months (2.5 years after randomisation), there was a significant improvement in all the efficacy endpoints (disease free survival, time to distant recurrence, time to contralateral breast cancer) in favour of exemestane, with a modest reduction in the risk of death[51]. The side effect profile of exemestane is similar to that of other aromatase inhibitors – with increased musculoskeletal complications and lower gynaecological side effects; and had no adverse effects on the patients' quality of life[51].

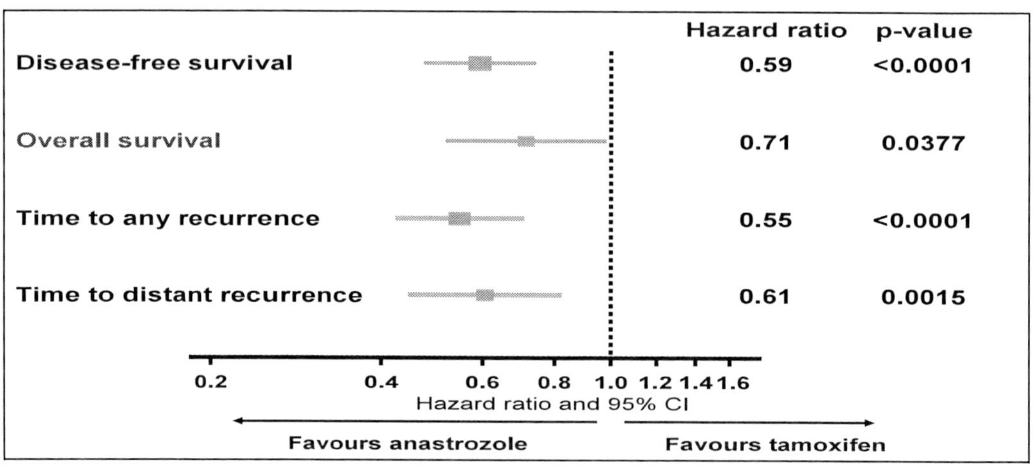

Abbreviations
ABCSG 8 = Austrian Breast and Colorectal cancer Study Group trial 8
ARNO 95 = Arimidex-Nolvadex 95 trial
ITA = Italian Tamoxifen Anastrozole.

Figure 9. Efficacy analysis of anastrozole vs tamoxifen in a meta-analysis of results from Austrian Breast & Colorectal Cancer Study group (ABCSG) 8/Arimidex-Nolvadex (ARNO) 95 and the Italian Tamoxifen Anastrozole (ITA) study.

'Extended' Adjuvant Therapy

Currently available evidence suggests that extending tamoxifen therapy beyond five years confers no additional survival benefit and may impose additional risks with side effects from treatment.[40, 52] Two further studies were designed to investigate whether using an aromatase inhibitor after a full five year course of tamoxifen can improve treatment outcomes. The National Cancer Institute of Canada Clinical Trials Group trial MA.17 randomly assigned 5187 postmenopausal, hormone-receptor-positive patients with early breast cancer who completed five years of tamoxifen to receive either letrozole or placebo. At the interim analysis of the trial (median follow up 2.4 years), patients treated with letrozole were found to have a significantly reduced risk of recurrence and improved disease free survival; with improved overall survival benefit in patients with node positive disease. This reached the predefined cutoff for the trial, and the data safety monitoring committee recommended for the treatment allocation to be unblinded, with women on the placebo arm being offered letrozole for a planned five years[53]. The outcome of this has recently been reported, with patients who have switched from placebo to letrozole having improved disease free survival (HR 0.37, $p<0.0001$) and distant disease free survival (HR 0.39; $p=0.004$)[54]. During intent-to-treat analysis of all outcomes, based on original randomisation to either placebo or letrozole arm, patients treated in the letrozole arm had a better outcome than the placebo group, despite 66% of placebo patients taking letrozole after unblinding. Four year disease free survival was significantly better in the letrozole arm (94.3% vs 91.4%, $p=0.0001$), with no significant benefit in distant disease free survival and overall survival[55].

In a separate study, the role of anastrozole in extended adjuvant therapy was investigated. In the Austrian Breast and Colorectal Cancer Study Group (ABCSG) Trial 6a, patients were randomised to receive three years of anastrozole or placebo, after remaining disease-free with five years of adjuvant tamoxifen. Patients on anastrozole had 38% reduced risk of recurrence (HR=0.63, $p=0.031$) with significantly fewer locoregional, contralateral or distant metastatic events. Despite that, there was no difference in overall survival in both arms[56].

Other Issues Concerning Adjuvant Use of Aromatase Inhibitors

As aromatase inhibitors are used more frequently in adjuvant endocrine therapy, their long term safety should be considered carefully. Though there have been fewer drug induced adverse events (ie. thromboembolic disease, hot flushes) and gynaecological side effects (ie. endometrial cancer, vaginal bleeding and discharge), there has been a report of increased vaginal dryness. This can lead to decreased libido, vaginal discomfort and dyspareunia[39] which can impact on their quality of life. This can usually be treated with vaginal moisturisers or lubricants.

A major adverse effect is the general increase in musculoskeletal complications (myalgia, arthralgia, carpal tunnel syndrome, paraesthesia) and fractures. Treatment with aromatase inhibitors can cause accelerated bone loss, which increases the risk of osteoporosis and fractures. Efforts have been made to investigate and manage this further, with the potential use of bisphosphonates 'prophylactically' in patients receiving adjuvant aromatase inhibitors[42, 57]. Arthralgia and myalgia are more frequent whilst treated on aromatase inhibitors. It is usually managed with oral analgesics, and do not seem overly troublesome to most patients, with low withdrawal rates due to arthralgia[58]. However, if the joint pain is severe and debilitating, switching to an alternative aromatase inhibitor or to tamoxifen, may improve the symptoms, and help maintain the benefit of aromatase inhibitor therapy[59].

Of concern as well is the increased cardiac events (myocardial infarction and cardiac failure) noted with letrozole and exemestane[44, 51]. This however, has not been shown in the ATAC trial with anastrozole (Table 7). Further analysis of the data revealed that any increased risk of myocardial infarction in the exemestane group was restricted to patients who were known hypertensive before beginning treatment. In the BIG 1-98 study, patients in the letrozole treatment group also had a significant increase in cardiovascular events (atherosclerosis, aneurysms, intermittent claudication) and low-grade cholesterol elevation, with a trend towards increased cardiac events. However, after adjusting for risk factors, this reached statistical significance only in patients aged 65-74, and was more marked in those with a previous history of hypertension or ischaemic heart disease[60].

As a result of these trials, aromatase inhibitors should be considered in adjuvant therapy for postmenopausal women with hormone receptor positive breast cancer. However, individual aromatase inhibitors are different, particularly with their safety profile, therefore the evidence for their use cannot be used interchangeably. All three third generation aromatase inhibitors (anastrozole, letrozole and exemestane) have been trialed, with proven efficacy as first line adjuvant therapy and switch therapy after two to three years of tamoxifen. Giving an aromatase inhibitor at the earliest opportunity is the best option to prevent recurrence in the first five years. However, with patients who are already on tamoxifen, they will benefit from switching to an aromatase inhibitor after two to three years. In those who are coming to the end of their 5 years of tamoxifen, extended adjuvant therapy with an aromatase inhibitors (letrozole or anastrozole based on data from available trails) should be considered, especially in patients with node-positive disease[54].

Future Directions

Premenopausal Women

Currently, aromatase inhibitors alone have no established role in premenopausal women with breast cancer. In premenopausal women, oestradiol levels are much higher than in postmenopausal women. Also, the relatively small effect of aromatase inhibitors on oestradiol levels will interrupt the negative feedback control of follicle-stimulating hormone (FSH) and luteinising hormone (LH) and cause a reciprocal increase in oestradiol secretion[61]. The increased oestrodiol levels can cause further tumour stimulation and follicular hyperplasia.

Table 7. Cardiovascular adverse events from the ATAC, BIG 1-98, MA 17, IES trials

		Aromatase Inhibitors (%)	Tamoxifen (%)	p
ATAC	Events	127(4.1)	104 (3.4)	0.12
	MI	37 (1.2)	34 (1.1)	
	Deaths	49 (1.4)	46 (1.4)	
BIG	Events	143 (3.6)	101 (2.5)	
	Deaths	26 (0.7)	13 (0.3)	
MA-17	Events	88 (4.1)	77 (3.6)	0.40
IES	MI	20 (0.9)	8 (0.4)	0.02*

* Statistically significant.

Abbreviations:

ATAC = Anastrozole, Tamoxifen, Alone or in Combination trial
BIG 1-98= Breast International Group 1-98 trial
MA 17 = National Cancer Institute of Canada clinical trials group trial MA 17
IES = International Exemestane Study
MI = myocardial infarction

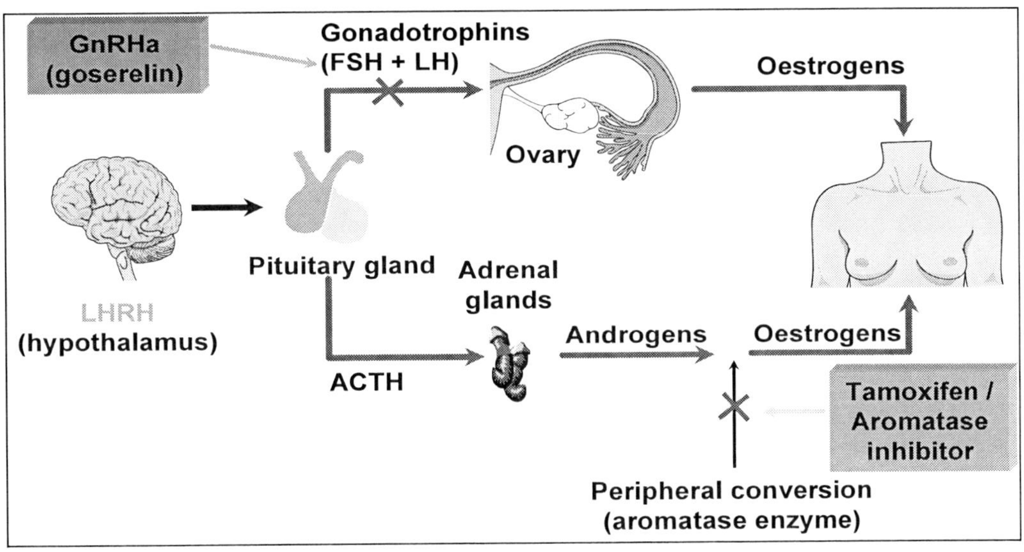

Abbreviations:

ACTH = Adrenocortocotrophic hormone
FSH = Follicle-stimulating hormone
LH = Luteinising hormone
GnRHa = Gonadotrophin-releasing hormone agonist

Figure 10. Rationale behind the use of aromatase inhibitors with ovarian suppression therapy in premenopausal women, highlighting point of action of GnRH agonist and endocrine therapy.

Table 8. Clinical response in a small study evaluating the use of goserelin with anastrozole as second-line therapy in premenopausal women with advanced breast cancer (who have progressed on tamoxifen)

N=16	n (%)
Partial response (PR)	1 (6)
Stable disease (SD) ≥6 months	9 (56)
Biochemical response* (BR)	2 (13)
Clinical benefit (PR + SD ≥6 months + BR) rate	75%

*no evidence of progression beyond 6 months with decreasing blood tumour markers

- Median duration of response 17+ months (range 6–47 months)

However, GnRH agonists (eg. goserelin) have been used in combination with tamoxifen in premenopausal women to inhibit both central and peripheral synthesis of oestradiol (Figure 10), with good results[62]. As an extension of this, recent studies have been carried out to investigate the combined use of anastrozole and goserelin in premenopausal women with advanced breast cancer. In patients who have progressed on goserelin and tamoxifen, conversion to anastrozole resulted in a further 76% reduction in mean oestradiol levels ($p<0.0001$). This translated into a significant clinical response of worthwhile duration (Table 8)[63]. When goserelin plus anastrozole was used as first line therapy in a similar group of patients, it produced a clinical benefit in a significant proportion of patients for a long duration[64, 65]. In oestrogen receptor positive premenopausal women with advanced breast cancer who have progressed on goserelin and a non-steroidal aromatase inhibitor (anastrozole) treatment, a small study showed that switching the patient to exemestane may offer some clinical benefit of reasonable duration. It also offers an extra therapeutic option in this particular group of patients[66].

The first efficacy results of a large study, the Austrian Breast and Colorectal Cancer Study Group trial 12 (ABCSG-12), evaluating the effect of adjuvant endocrine therapy with ovarian suppresion, with or without the bisphosphonate zoledronic acid in premenopausal women with hormone responsive tumours have recently been reported. There was no significant difference in disease free survival in patients treated with adjuvant endocrine therapy (tamoxifen vs anastrozole) with ovarian suppression alone. However, the addition of zoledronic acid significantly prolonged disease free survival and relapse free survival compared with adjuvant endocrine therapy alone (with ovarian suppression) in premenopausal women[67]. A further substudy on the effect of zoledronic acid on bone mineral density showed that premenopausal patients on endocrine therapy alone suffered significant bone loss; whilst zoledronic acid prevented bone loss and improved bone mineral density at five years[68].

Some breast cancer units have also used aromatase inhibitors alone in selected premenopausal patients with chemotherapy induced amenorrhoea (on the premise that they have chemotherapy induced ovarian failure – hence 'postmenopausal'). However, this should be done with caution as there have been reports of return of ovarian function, which could lead to premenopausal oestradiol levels, with the loss of anticancer effect of aromatase inhibitors and ovulation, with the risk of pregnancy[69].

Currently, there is no conclusive evidence to suggest that aromatase inhibitors with ovarian suppression is superior to tamoxifen with ovarian suppression as adjuvant therapy in premenopausal women. However, there has been evidence that they have a role to play in premenopausal women with oestrogen receptor positive advanced breast cancer.

Neoadjuvant Therapy

Neoadjuvant systemic therapy appears to have the potential of down staging a primary breast cancer, which enables more women to have breast conserving surgery who would have otherwise required a mastectomy[70]. It also allows for assessment of therapeutic response, hence better targeting of later therapy after surgery. On the other hand, as will be explained later, there is yet no data available in relation to local recurrence and overall survival for patients treated this way.

Table 9. Results summary of a randomised controlled trial evaluating the use of letrozole compared with tamoxifen as neoadjuvant therapy in postmenopausal women with oestrogen receptor positive primary breast cancer

	Letrozole n=154	Tamoxifen n=170	P-value
Objective Response			
Clinical	**55%**	**36%**	**<0.001**
Ultrasound	**35%**	**25%**	**0.042**
Mammography	**34%**	**16%**	**<0.001**
Breast-Conserving Surgery	**45%**	**35%**	**0.022**

Preoperative endocrine therapy with tamoxifen has been reported to produce a response rate approaching 70% in 3 months with oestrogen receptor positive tumours[71]. Endocrine therapy has advantages over chemotherapy in having fewer side effects, and that it can be continued throughout the perioperative period. As aromatase inhibitors offers good clinical benefit with a better side effect profile over tamoxifen (eg. lower thromboembolic risk, which is relevant during the perioperative period), their use in neoadjuvant therapy have been investigated.

Eiermann and colleagues have investigated the use of neoadjuvant letrozole compared with tamoxifen in oestrogen receptor positive disease before definitive surgery. All the women recruited into the study were ineligible for breast-conserving surgery, with 14% considered inoperable. Patients treated in the letrozole arm had a significantly better objective response rate, whether measured by clinical palpation, mammogram or ultrasound. This led to a more letrozole-treated patients being eligible for breast conserving surgery (Table 9) [72]. As a result, letrozole is currently licensed for use in the preoperative setting.

The Immediate Preoperative Anastrozole, Tamoxifen, or Combined with Tamoxifen (IMPACT) trial randomised 330 postmenopausal women with oestrogen receptor positive tumours to receive either tamoxifen, anastrozole or both tamoxifen and anastrozole for 3 months before surgery. When comparing clinical objective response rates, there was no significant difference between patients in the different treatment groups. However, a significant number of women(46%) were deemed eligible for breast conserving surgery, who previously required mastectomy, after treatment with anastrozole (22 % with tamoxifen, $p=0.03$)[73]. Semiglazov et al. compared the use of exemestane against tamoxifen as neoadjuvant therapy in postmenopausal patients with oestrogen receptor positive tumours. Patients receiving exemestane had a significant improvement in clinical objective response (76% vs 40%, $p=0.05$), with higher rates of breast conserving surgery (37% vs 20%, $p=0.05$)[74].

These trials add to the evidence that third generation aromatase inhibitors are significantly more effective than tamoxifen in downstaging large tumours, hence reducing the need for mastectomy in postmenopausal oestrogen receptor positive women. However, further studies are required to evaluate the long term clinical outcome of patients treated with neoadjuvant therapy and compare them with patients who were treated with 'conventional' adjuvant therapy; particularly addressing concerns regarding rates of local recurrence, disease free survival and overall survival. Studies are also required to identify factors, in addition to oestrogen receptor status, to select patients who would benefit most from neoadjuvant endocrine therapy.

Primary Therapy for Early Breast Cancer in the Elderly

In elderly patients with early primary breast cancer, the initial treatment option is often surgery. However, some patients may prefer not to have surgery or are medically unfit for surgery. Previous studies have evaluated the use of tamoxifen in this context[71, 75], but little is known about the efficacy of aromatase inhibitors in this setting. A recent small series reported that 93.8% of patients treated with anastrozole reached clinical benefit at 6 months[76]. Currently, a large multicentre clinical trial (Endocrine or Surgical Therapy for Elderly women with Mammary cancer (ESTEeM)) is ongoing in the UK to compare the efficacy of surgery (with adjuvant anastrozole treatment) against anastrozole as primary treatment in women over 75 years with oestrogen receptor positive invasive breast cancer.

Chemoprevention

Tamoxifen has been shown to have a potential role in chemoprevention, with reduction of oestrogen receptor positive breast cancers in 30-40% of high risk women for up to 10 years[77-80]. As such, tamoxifen has been licensed for chemoprevention use in United States. As third generation aromatase inhibitors have been shown to be more effective than tamoxifen in other treatment settings (adjuvant and neoadjuvant therapy in early and advanced breast cancer), they may potentially be more effective than tamoxifen in the chemoprevention setting, with a better side effect profile. Analysis of completed randomised controlled trials comparing third generation aromatase inhibitors against tamoxifen shows a 53% reduction in all contralateral breast cancers in the aromatase inhibitor group and a 57% reduction in invasive contralateral breast cancer in oestrogen receptor positive patients (Figure 11)[36, 38, 44, 51, 53, 55].

The international multi-centre study – International Breast Cancer Intervention Study(IBIS) II (Prevention) is ongoing, comparing five years of anastrozole treatment with placebo in postmenopausal women at increased risk of breast cancer. The primary end point is the development of breast cancer, with a secondary endpoint of breast cancer mortality. Throughout the study, major efforts are also taken to monitor the effect of anastrozole on bone mineral density with bisphosphonate treatment for patients at an increased risk of osteoporosis and fractures[81].

Biological Modulation

The improved efficacy of aromatase inhibitors over tamoxifen, may be related to the partial oestrogen agonistic properties of tamoxifen which may limit its efficacy in breast cancer[82]. Previous laboratory studies have suggested that different molecular characteristics that vary between cell lines and breast cancer tumours, may affect the efficacy of endocrine therapy in breast cancer. Expression of a progresterone receptor, a key molecular marker for breast cancer, has been found to be highly oestrogen dependent, where its expression is indicative of an intact oestrogen receptor mechanism. To date, there has been minimal data on specific molecular markers in breast cancer and their relation to endocrine therapy and clinical outcome. Analysis of these molecular features may identify subpopulations which may derive greater benefit from a particular endocrine therapy. This will improve understanding of the molecular characteristics of breast cancer and may allow for better patient selection for targeted therapies.

In the ATAC study, a retrospective analysis was carried out to evaluate the effect of progesterone receptor status on time to recurrence. Initial data suggested that patients in the oestrogen receptor positive / progesterone receptor negative subgroup who were treated with anastrozole had only half the number of recurrences compared to patients treated with tamoxifen. This is significantly different to that in patients with oestrogen receptor positive / progesterone receptor positive subgroup, where they had approximately 16% lower recurrence rate when treated with anastrozole relative to tamoxifen[83]. However, further centralised re-analysis of hormone receptor status on the original tumour blocks of the patients has failed to confirm this. In this re-analysis, quantitative expression of oestrogen receptor or progesterone receptor did not provide any differential relative benefit from anastrozole over tamoxifen (Figure 12). Patients treated with anastrozole had an improved time to recurrence compared with tamoxifen, in all molecular subgroups[84]. The cause of this difference is unclear, but the earlier results may be an artifact of unplanned subgroup analysis.

The results of the recent subgroup analysis of ATAC patients is consistent with the subgroup analysis within the BIG 1-98 trial, which showed no significant difference between the oestrogen receptor positive / progesterone receptor negative and oestrogen receptor positive / progesterone receptor positive subgroup in disease free survival with letrozole[44]. However, these results should be interpreted with caution as neither the ATAC trial nor the BIG 1-98 trial were powered to detect differences via retrospective analyses of sub-populations and any apparent differences should not be relied upon to make clinical decisions.

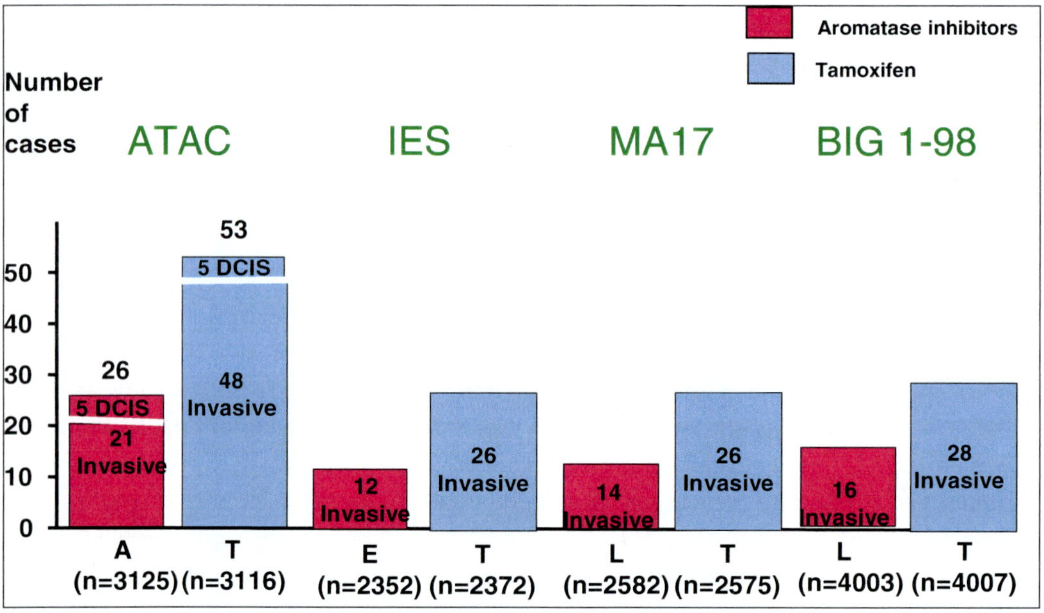

Abbreviations:

ATAC = Anastrozole, Tamoxifen, Alone or in Combination trial
IES = International Exemestane Study
NCIC MA 17 = National Cancer Institute of Canada clinical trials group trial MA 17
BIG 1-98 = Breast International Group 1-98 trial
DCIS = Ductal carcinoma in-situ
A = Anastrozole
T = Tamoxifen
E = Exemestane
L = Letrozole

Figure 11. Combined analysis of completed randomised controlled trials of aromatase inhibitors vs tamoxifen as adjuvant therapy, measuring the incidence of new (contralateral) primary breast cancer.

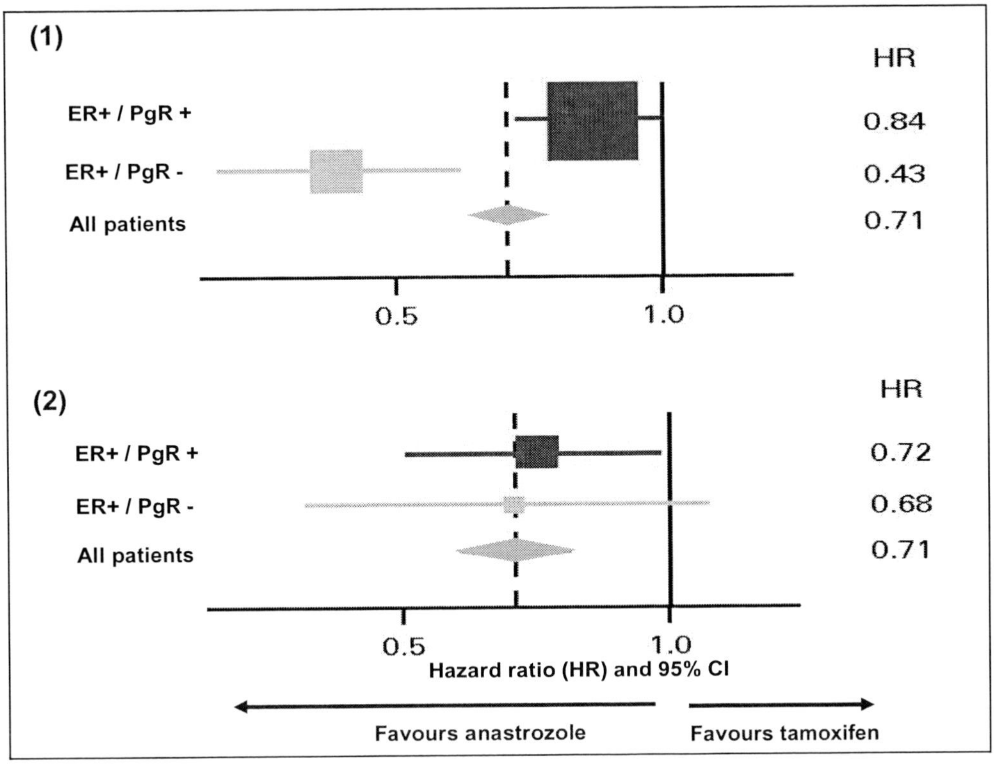

Abbreviations:
ER = oestrogen receptor
PgR = progesterone receptor

Figure 12. Time to recurrence for patients treated with anastrozole vs tamoxifen from the subgroup analysis of Anastrozole, Tamoxifen, Alone or in Combination (ATAC) trial, with patients classified by oestrogen receptor and progesterone receptor status – (1) original analysis of ATAC patients with results determined by local laboratory; (2) results from centralised re-analysis of hormone receptor status.

A recent study reported that increased progesterone receptor levels in breast cancer tumour tissue was associated with increased time to treatment failure with endocrine therapy, in oestrogen receptor positive advanced breast cancer[85]. Further studies are required to correlate these findings and identify other molecular and biological factors that may influence the response of a tumour to endocrine therapy. Also prospective randomised controlled trials need to be carried out to evaluate the efficacy of aromatase inhibitors in specific groups of patients with different biological markers, measuring long term outcomes. This may help refine the selection of patients for endocrine therapy to achieve the maximal response.

Combination Therapies

There has not yet been sufficient evidence to suggest that combination therapy with two endocrine therapy agents confer a significant advantage over monotherapy. However, newer, more potent agents have been developed aiming at alternative 'targets' in breast cancer tumour cells - Epidermal Growth Factor Receptor(EGFR) and Human Epidermal growth factor Receptor 2 (HER2), which could potentially be used in combination with endocrine therapy to improve its efficacy.

A potential 'target' that is being investigated is EGFR, whose over-expression has been associated with tamoxifen resistance, with 'cross-talks' between oestrogen receptor and EGFR pathways[86]. Preclinical data has suggested that EGFR could play a role in overcoming endocrine therapy resistance in breast cancer. A study comparing treatment with gefitinib (an EGFR tyrosine kinase inhibitor) in combination with anastrozole, against anastrozole alone was presented recently. The trial was closed prematurely due to poor patient recruitment. However, analysis of results (n=71) showed that the addition of gefitinib did not delay the onset of resistance to endocrine therapy (anastrozole). Also, toxicity and adverse events with gefitinib led to cessation of therapy in nearly one-third of patients[87].

HER2 is a transmembrane glycoprotein receptor that is overexpressed in 13-30% of human breast cancers. It is thought to be associated with aggressive pathology and poor prognosis in metastatic breast cancer. It has also been associated with resistance to endocrine therapy. Trastuzumab is a recombinant, humanised HER2 monoclonal antibody, which has been shown to be effective in metastatic breast cancer[88]. Several trials have evaluated the use of trastuzumab in combination with aromatase inhibitors in advanced breast cancer. A small study evaluating the use of trastuzumab with letrozole showed an overall response rate of 26%, with clinical benefit rate of 52% in women with hormone receptor positive and HER2 overexpressing advanced breast cancer[89]. Another recent trial comparing the use of trastuzumab in combination with anastrozole against anastrozole alone showed a significant improvement in median progression free survival in the combination arm (4.8 months vs 2.4 months, p=0.0016) in hormone-dependent, HER2 overexpressing metastatic breast cancer women. However, there was no significant difference in overall survival[88]. Several studies are ongoing evaluating the use of trastuzumab in combination with other aromatase inhibitors (letrozole or exemestane) in hormone-dependent, HER2 positive, postmenopausal women with advanced breast cancer.

The combination of trastuzumab and letrozole treatment was reasonably well tolerated, with the majority being grade one or two side effects that were consistent with those known to be associated with letrozole or trastuzumab monotherapy. Fatigue, hot flushes, arthralgia and pain were the more commonly reported side effects. However, trastuzumab has been associated with an increased risk of cardiac dysfunction and can cause cardiomyopathy, precipitating congestive cardiac failure[90]. This needs to be taken into consideration when using trastuzumab, balancing the potential benefit of breast cancer treatment with the adverse effect of cardiac toxicity.

Lapatinib is an oral dual tyrosine-kinase inhibitor of HER2 and EGFR pathways. Preclinical studies showed evidence of synergy between lapatinib and tamoxifen in models of endocrine resistance. As such, a large randomised controlled phase III trial was designed to compare the use of lapatinib and letrozole against letrozole alone in oestrogen receptor positive postmenopausal women with advanced metastatic breast cancer. The first results of the study were presented recently, which showed that lapatinib significantly improved the clinical efficacy of an aromatase inhibitor. Women who are oestrogen receptor positive and HER2 positive treated in the combination arm had significantly better overall response rate (28% vs 15%, p=0.021), better clinical benefit rate (48% vs 29%, p=0.003) with a 29%

decrease in risk of disease progression (p=0.019) and an improvement in median progression free survival from 3.0 to 8.2 months. However, there was no significant difference in overall survival at 45 months. This benefit was not reciprocated in the HER2 negative population group. The combination of letrozole and lapatinib had a manageable and predictable side effect profile, with no unexpected toxicity. As such, lapatinib and letrozole in combination offer a good first line oral treatment approach for post menopausal women with oestrogen receptor positive, HER2 positive metastatic breast cancer[91].

Long term oestrogen deprivation in human breast cancer cells could lead to upregulation of oestrogen receptor expression[92], which in turn causes cell proliferation. Different endocrine therapy agents produce different degrees of inhibition with different oestradiol levels(Figure 13), where fulvestrant (Faslodex® - a pure anti-oestrogen and a selective oestrogen receptor downregulator) is thought to cause maximal inhibition. Fulvestrant has also been shown to inhibit the growth of long term oestrogen deprived cells, which are resistant to tamoxifen[93]. This may have therapeutic implications in patients who have relapsed on tamoxifen or aromatase inhibitors. As such, various trials are currently ongoing which evaluates the use of fulvestrant combined with a third generation aromatase inhibitor.

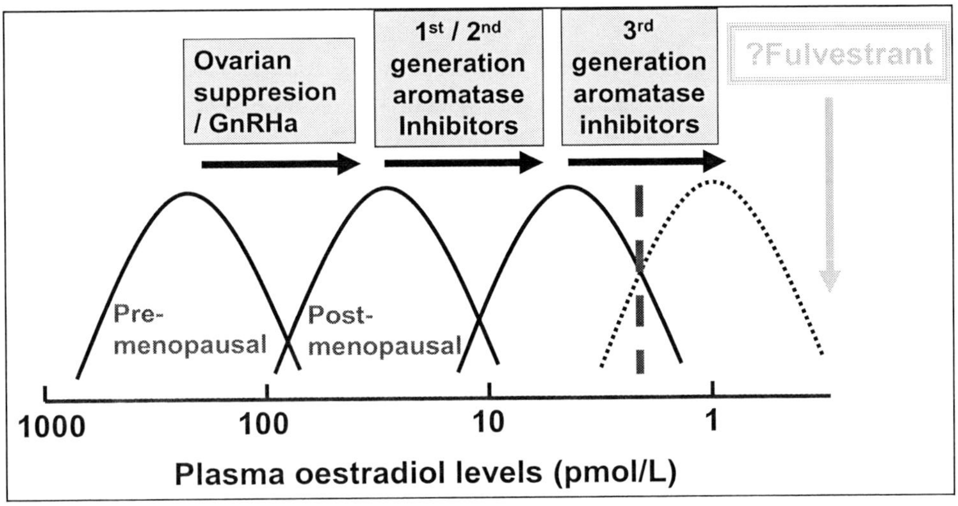

Abbreviations:
GnRHa = gonadotrophin-releasing hormone agonist.

Figure 13. Changes in plasma oestradiol levels with oestrogen deprivation using different endocrine therapy agents.

The Study of Faslodex with or without Concomitant Arimidex vs Exemestane Following Progression on Non-steroidal Aromatase Inhibitors (SoFEA) is currently recruiting oestrogen receptor positive postmenopausal patients with advanced breast cancer who have progressed on nonsteroidal aromatase inhibitors to be randomised to receive either 'fulvestrant + anastrozole', fulvestrant alone or exemestane, with the primary endpoint being time to progression. The Faslodex and Arimidex Clinical Trial (FACT) trial compares the use of 'anastrazole + fulvestrant' with anastrozole alone, in postmenopausal patients with oestrogen receptor positive breast cancer in first relapse after primary treatment of localised tumour. The results of these trials will further evaluate the potential role of aromatase inhibitors in combination therapy for postmenopausal women with oestrogen receptor positive breast cancer.

The studies evaluating the use of aromatase inhibitors in combination therapy (with pure antioestrogens or biological therapy) has showed some promising results. Further studies are currently ongoing to evaluate this further and to identify the most suitable combination of treatment for more targeted therapy in patients with advanced breast cancer.

Conclusion

Aromatase inhibitors clearly have an established role in endocrine therapy for breast cancer, particularly in postmenopausal hormone receptor positive disease. Third generation aromatase inhibitors are very well tolerated, with potentially manageable side effects. They have an established role in the treatment of advanced breast cancer, and as an adjuvant therapy in early primary breast cancer. Studies of the use of aromatase inhibitors in premenopausal women (alongside ovarian suppression therapy) and neoadjuvant treatment have shown promising results. The potential role of aromatase inhibitors in chemoprevention of breast cancer in high risk women or as combination therapies (with biological therapy and pure antioestrogens) are the subject of ongoing trials. Future developments in biological modulation may lend itself to better patient selection and targeted therapy with aromatase inhibitors.

References

[1] Beatson, G., On the treatment of inoperable cases of carcinoma of the mamma: suggestion for a new method of treatment, with illustrative cases. *Lancet,* 1896. 2: p. 104-107.

[2] Brueggemeier, R., J. Hackett, and E. Diaz-Cruz, Aromatase Inhibitors in the treatment of breast cancer. *Endocrine reviews,* 2005. 26(3): p. 331-345.

[3] Hawkins, R., et al., Prognostic significance of oestrogen and progestogen receptor activities in breast cancer. *British Journal of Surgery,* 1987. 74(11): p. 1009-1113.

[4] Knight, W.A., et al., Estrogen receptor as an independent prognostic factor for early recurrence in breast cancer. *Cancer Research,* 1977. 37: p. 4669-4671.

[5] Maynard, P., et al., Estrogen receptor assay in primary breast cancer and early recurrence f the disease. *Cancer Research,* 1978. 38: p. 4292-4295.

[6] Osborne, C., et al., The value of oestrogen and progesterone receptors in the treatment of breast cancer. *Cancer,* 1980. 46(12): p. 2884-2888.

[7] Thorpe, S., Estrogen and progesterone receptor determinations in breast cancer. Technology, biology and clinical significance. *Acta Oncologica,* 1988. 27(1): p. 1-19.

[8] Cheung, K., et al., Pathological features of primary breast cancer in the elderly based on needle core biopsies - a large series from a single centre. *Critical Reviews in Oncology/Haematology,* 2008. 67(3): p. 263-267.

[9] Diab, S., R. Elledge, and G. Clark, Tumor Characteristics and clinical outcome of elderly women wtih breast cancer. *Journal of National Cancer Institute,* 2000. 92(7): p. 550-556.

[10] Allegra, J., et al., Changes in multiple or sequential estrogen receptor determinates in breast cancer. *Cancer,* 1980. 45(4): p. 792-794.

[11] Gapinski, P. and W. Donegan, Estrogen receptors and breast cancer: prognostic and therapeutic implications. *Surgery,* 1980. 88(3): p. 386-393.

[12] Samaan, N., et al., Estrogen receptor: a prognostic factor in breast cancer. *Cancer,* 1981. 47(3): p. 554-560.

[13] Williams, M., et al., Oestrogen receptors in primary and advanced breast cancer: an eight year review of 704 cases. *British Journal of cancer,* 1987. 55(1): p. 67-73.

[14] Cole, M., C. Jones, and I. Todd, A new antioestrogenic agent in late breast cancer. An early appraisal of ICI 46,474. *British Journal of cancer,* 1971. 25: p. 270-275.

[15] Bickenbach, K. and N. Jaskowiak, Aromatase Inhibitors: an overview for surgeons. *Journal of American College of Surgeons,* 2006. 203(3): p. 376-389.

[16] James, V., et al., Aromatase activity in normal breast and breast tumor tissues: in vivo and in vitro studies. *Steroids,* 1987. 50: p. 269-279.

[17] Miller, W., Aromatase activity in breast tissue. *Journal of Steroid Biochemistry Molecular Biology,* 1991. 39: p. 783-790.

[18] Santner, S., et al., Aromatase activity and expression in breast cancer and benign breast tissue stromal cells. *Journal of Clinical Endocrinology & Metabolism,* 1997. 82(1): p. 200-208.

[19] Cocconi, G., First generation aromatase inhibitors - aminoglutethimide and testololactone. *Breast Cancer Research and Treatment,* 1994. 30(1): p. 57-80.

[20] Lipton, A. and R. Santen, Proceedings: medical adrenalectomy using aminoglutethimide and dexamethasone in advanced breast cancer. *Cancer,* 1974. 33: p. 503-512.

[21] Santen, R., et al., Kinetic, hormonal and clinical studies with aminoglutethimide in breast cancer. *Cancer,* 1977. 39(2948-2958).

[22] Brodie, A., Aromatase inhibitors in breast cancer. *Trends in endocrinology & Metabolism,* 2002. 13(2): p. 61-65.

[23] Dowsett, M., et al., In vivo measurement of aromatase inhibition by letrozole (CGS 20267) in postmenopausal patients with breast cancer. *Clinical Cancer Research,* 1995. 1: p. 1511-1515.

[24] Evans, T., et al., Phase I and endocrine study of exemestane (FCE 24304), a new aromatase inhibitor, in postmenopausal women. *Cancer Research,* 1992. 52(5933-5939).

[25] Geisler, J., et al., Influence of anastrozole (Arimidex), a selective, non-steroidal aromatase inhibitor, on in vivo aromatization and plasma oestrogen levels in postmenopausal women with breast cancer. *British Journal of Cancer,* 1996. 74(1286-1291).

[26] Demers, L., Effects of fadrozole (CGS 16949A) and letrozole (CGS 20267) on the inhibition of aromatase activity in breast cancer patients. *Breast Cancer Research and Treatment,* 1994. 30: p. 95-102.

[27] Geisler, J., et al., Influence of letrozole and anastrozole on total body aromatization and plasma estrogen levels in postmenopausal breast cancer patients evaluated in a randomized cross-over study. *Journal of Clinical Oncology,* 2002. 20(3): p. 751-757.

[28] Buzdar, A., et al., Anastrozole, a potent and selective aromatase inhibitor, versus megestrol acetate in postmenopausal women with advanced breast cancer: results of overview analysis of two phase III trials. Arimidex Study Group. *Journal of Clinical Oncology,* 1996. 14(7): p. 2000-2011.

[29] Buzdar, A., et al., Anastrozole versus megestrol acetate in the treatment of postmenopausal women with advanced breast carcinoma: results of a survival update based on a combined analysis of data from two mature phase III trials. Arimidex study group. *Cancer,* 1998. 83(6): p. 1142-1152.

[30] Dombernowsky, P., et al., Letrozole, a new oral aromatase inhibitor for advanced breast cancer: double-blind randomized trial showing a dose effect and improved efficacy and tolerability compared with megestrol acetate. *Journal of Clinical Oncology,* 1998. 16(2): p. 453-461.

[31] Kaufmann, M., et al., Exemestane is superior to megestrol acetate after tamoxifen failure in postmenopausal women with advanced breast cancer: results of a phase III randomized double-blind trial. The Exemestane Study Group. *Journal of Clinical Oncology,* 2000. 18(7): p. 1399-1411.

[32] Nabholtz, J., et al., Anastrozole is superior to tamoxifen as first line therapy for advanced breast cancer in postmenopausal women: results of a North American multicentre randomized trial. Arimidex study group. *Journal of Clinical Oncology,* 2000. 18(22): p. 3758-3767.

[33] Bonneterre, J., et al., Anastrozole versus tamoxifen as first-line therapy for advanced breast cancer in 668 postmenopausal women: results of the tamoxifen or arimidex randomized group efficacy and tolerability study. *Journal of Clinical Oncology,* 2000. 18(22): p. 3748-3757.

[34] Thurlimann, B., et al., Efficacy of tamoxifen following anastrozole ('Arimidex') compared with anastrozole following tamoxifen as first-line treatment for advanced breast cancer in postmenopausal women. *European Journal of Cancer,* 2003. 39(16): p. 2310-2317.

[35] Mouridsen, H., et al., Superior efficacy of letrozole versus tamoxifen as first line therapy for postmenopausal women with advanced breast cancer: results of a phase III study of the International Letrozole Breast Cancer Group. *Journal of Clinical Oncology,* 2001. 19(10): p. 2596 - 2606.

[36] Paridaens, R., et al., Phase III study comparing exemestane with tamoxifen as first-line hormonal treatment of metastatic breast cancer in postmenopausal women: the European Organisation for Research and Treatment of Cancer Breast Cancer Cooperative Group. *Journal of Clinical Oncology,* 2008. 26(30): p. 4883-4890.

[37] Saphner, T., D. Tormey, and R. Gray, Annual hazard rates of recurrence for breast cancer after primary therapy. Journal of Clinical Oncology, 1996. 14: p. 2738-2746.

[38] Howell, A., et al., Results of the ATAC (Arimidex, Tamoxifen, Alone or in Combination) trial after completion of 5 years adjuvant treatment for breast cancer. *Lancet,* 2005. 365(9453): p. 60-62.

[39] The Arimidex, T., Alone or in combination(ATAC) trialists' group, Effect of anastrozole and tamoxifen as adjuvant treatment for early-stage breast cancer: 100-month analysis of the ATAC trial. *Lancet Oncology,* 2008. 9(45-53).

[40] Group, E.B.C.T.C., Tamoxifen for early breast cancer: an overview of the randomised trials. *Lancet,* 1998. 351(9114): p. 1451-1467.

[41] Eastell, R., et al., Effect of anastrozole on bone mineral density: 5-year results from the anastrozole, tamoxifen, alone or in combination trial 18233230. *Journal of Clinical Oncology,* 2008. 26(7): p. 1051-1057.

[42] Aapro, M., Improving bone health in patients with early breast cancer by adding bisphosphonates to letrozole: the Z-ZO-E-ZO-FAST program. *Breast,* 2006. 15: p. 30-40.

[43] Brufsky, A., et al., Zoledronic acid inhibits adjuvant letrozole-induced bone loss in postmenopausal women with early breast cancer. *Journal of Clinical Oncology,* 2007. 25(7): p. 829-836.

[44] Coates, A., et al., Five years of letrozole compared with Tamoxifen as initial adjuvant therapy for postmenopausal women with Endocrine-responsive early breast cancer: update of study BIG 1-98. *Journal of Clinical Oncology,* 2007. 25(5): p. 486-492.

[45] Mouridsen, H., et al., BIG 1-98: A randomized double-blind phase III study evaluating letrozole and tamoxifen given in sequence as adjuvant endocrine therapy for postmenopausal women with receptor-positive breast cancer. *Cancer Research,* 2008. 69(Suppl.)(2): p. 13.

[46] Jones, S., et al., Results of the first planned analysis of the TEAM (tamoxifen exemestane adjuvant multinational) prospective randomized phase III trial in hormone sensitive postmenopausal early breast cancer. *Cancer Research,* 2008. 69(Suppl.)(2): p. 15.

[47] Jakesz, R., et al., Tamoxifen and anastrozole as a sequencing strategy in postmenopausal women with hormone-responsive early breast cancer: updated data from the Austrian breast and colorectal cancer study group trial 8. *Cancer Research,* 2008. 69(Suppl.)(2): p. 14.

[48] Boccardo, F., et al., Switching to anastrozole versus continued tamoxifen treatment of early breast cancer. Updated results of the Italian tamoxifen anastrozole (ITA) trial. *Annals of Oncology,* 2006. 17(Supp 7): p. 10-14.

[49] Jakesz, R., et al., Switching of postmenopausal women with endocrine-responsive early breast cancer to anastrozole after 2 years' adjuvant tamoxifen: combined results of ABCSG trial 8 and ARNO 95 trial. *Lancet,* 2005. 366(9484): p. 455-462.

[50] Jonat, W., et al., Effectiveness of switching from adjuvant tamoxifen to anastrozole in postmenopausal women with hormone-sensitive early-stage breast cancer: a meta-analysis. *The Lancet Oncology,* 2006. 7(12): p. 991-996.

[51] Coombes, R., et al., Survival and safety of exemestane versus tamoxifen after 2-3 years' tamoxifen treatment (Intergroup Exemestane Study): a randomised controlled trial. *Lancet,* 2007. 369(9561): p. 559-570.

[52] Fisher, B., et al., Five versus more than five years of tamoxifen for lymph node-negative breast cancer: updated findings from the National Surgical Adjuvant breast

and Bowel Project B-14 randomized trial. *Journal of National Cancer Institute,* 2001. 93(9): p. 684-690.

[53] Goss, P., et al., A Randomized trial of Letrozole in postmenopausal women afer five years of tamoxifen therapy for early-stage breast cancer. *New England Journal of Medicine,* 2003. 349(19): p. 1793-1802.

[54] Goss, P., et al., Late extended adjuvant treatment with letrozole improves outcome in women with early-stage breast cancer who complete 5 years of tamoxifen. *Journal of Clinical Oncology,* 2008. 26(12): p. 1948-1955.

[55] Ingle, J., et al., Intent-to-treat analysis of the placebo-controlled trial of letrozole for extended adjuvant therapy in early breast cancer: NCIC CTG MA.17. *Annals of Oncology,* 2008. 19(5): p. 877-882.

[56] Jakesz, R., et al., Extended adjuvant therapy with anastrozole among postmenopausal breast cancer patients: results from the randomized Austrian Breast and Colorectal cancer Study Group Trial 6a. *Journal of National Cancer Institute,* 2007. 99(24): p. 1845-1853.

[57] Eidtmann, H., et al., The effect of zoledronic acid on aromatase inhibitor associated bone loss in postmenopausal women with early breast cancer receiving adjuvant letrozole: 36 months follow-up of ZO-FAST. *Cancer Research,* 2008. 69(Suppl.)(2): p. 44.

[58] The Arimidex, T., Alone or in combination(ATAC) trialists' group, Comprehensive side-effect profile of anastrozole and tamoxifen as adjuvant treatment for early-stage breast cancer: long term safety analysis of the ATAC trial. *Lancet Oncology,* 2006. 7: p. 633-643.

[59] Briot, K., et al., Effects of switching aromatase inhibitors on arthralgia: the ATOLL study. *Cancer Research,* 2008. 69(Suppl.)(2): p. 1142.

[60] Crivellari, D., et al., Letrozole compared with tamoxifen for elderly patients with endocrine-responsive early breast cancer: the BIG 1-98 Trial. *Journal of Clinical Oncology,* 2008. 26(12): p. 1972-1979.

[61] Tredway, D., et al., A phase I study of the pharmacokinetics, pharmacodynamics and safety of single and multiple-dose anastrozole in healthy, premenopausal female volunteers. *Fertil Steril,* 2004. 82(6): p. 1587-1593.

[62] Klijn, J., et al., Combined tamoxifen and luteinizing hormone-releasing hormone (LHRH) agonist versus LHRH agonist alone in premenopausal advanced breast cancer: a meta-analysis of four randomized trials. *Journal of Clinical Oncology,* 2001. 19: p. 343-353.

[63] Forward, D., et al., Clinical and endocrine data for goserelin plus anastrozole as second-line endocrine therapy for premenopausal advanced breast cancer. *British Journal of cancer,* 2004. 90(3): p. 590-594.

[64] Cheung, K., et al., Goserelin plus anastrozole as first-line endocrine therapy for premenopausal women with oestrogen receptor(ER) positive advanced breast cancer (ABC). *Journal of Clinical Oncology,* 2005. 23(16S): p. 731.

[65] Carlson, R., et al., Goserelin plus anastrozole in the treatment of premenopausal hormone receptor positive, recurrent or metastatic breast cancer. *Journal of Clinical Oncology,* 2007. 25(18S): p. 1030.

[66] Agrawal, A., et al., Continued use of goserelin to achieve ovarian function suppression in combination with a further aromatase inhibitor (exemestane) following prior treatment with anastrozole and/or tamoxifen in premenopausal women with oestrogen receptor-positive advanced breast cancer. *European Journal of Cancer,* 2006. 4(Supp 1): p. 409.

[67] Gnant, M., et al., Adjuvant ovarian suppression combined with tamoxifen or anastrozole, alone or in combination with zoledronic acid, in premenopausal women with hormone-responsive, stage I and II breast cancer: First efficacy results from ABCSG-12. *Journal of Clinical Oncology,* 2008. 26((May 20 Suppl; abstr LBA4)).

[68] Gnant, M., et al., Adjuvant endocrine therapy plus zoledronic acid in premenopausal women with early-stage breast cancer: 5-year follow-up of the ABCSG-12 bone-mineral density substudy. *The Lancet Oncology,* 2008. 9(9): p. 840-849.

[69] Smith, I., et al., Adjuvant Aromatase Inhibitors for Early Breast Cancer after Chemotherapy-Induced Amenorrhoea: Caution and Suggested Guidelines. *Journal of Clinical Oncology,* 2006. 24(16): p. 2444-2447.

[70] Wolmark, N., et al., Preoperative chemotherapy in patients with operable breast cancer: nine-year results from National Surgical Adjuvant Breast and Bowel Project B-18. *Journal of National Cancer Institute Monographs,* 2001. 30: p. 96-102.

[71] Cheung, K., A. Howell, and J. Robertson, Preoperative endocrine therapy for breast cancer. *Endocrine-Related Cancer,* 2000. 7: p. 131-141.

[72] Eiermann, W., et al., Preoperative treatment of postmenopausal breast cancer patients with letrozole: A randomized double-blind multicentre study. *Annals of Oncology,* 2001. 12(11): p. 1527-1532.

[73] Smith, I., et al., Neoadjuvant treatment of postmenopausal breast cancer with Anastrozole, Tamoxifen or both in combination: The Immediate Preoperative Anastrozole, Tamoxifen or combined with Tamoxifen (IMPACT) Multicenter Double-Blind Randomized Trial. *Journal of Clinical Oncology,* 2005. 23(22): p. 5108-5116.

[74] Semiglazov, V., et al., Exemestane (E) vs Tamoxifen (T) as neoadjuvant endocrine therapy for postmenopausal women with ER+ breast cancer (T2N1-2, T3N0-1, T4N0M0). *Journal of Clinical Oncology,* 2005. 23(16S): p. 530.

[75] Hind, D., et al., Surgery versus primary endocrine therapy for operable primary breast cancer in elderly women (70 years plus) (Review). *The Cochrane Library,* 2006(3).

[76] Cheung, K., et al., Primary endocrine therapy with anastrozole for early operable primary breast cancer in the elderly: Early clinical and biological data. *Journal of Clinical Oncology,* 2006. 24(18S): p. 10647.

[77] Cuzick, J., et al., Long-term results of tamoxifen prophylaxis for breast cancer - 96 month follow up of the randomised IBIS-I trial. *Journal of National Cancer Institute,* 2007. 99(4): p. 272-282.

[78] Fisher, B., Tamoxifen for prevention of breast cancer: report of the National Surgical Adjuvant Breast and Bowel Project P-1 study. *Journal of National Cancer Institute,* 1998. 90(1371-1388).

[79] Ross, P. and T. Powles, Results and implications of the Royal Marsden and other tamoxifen chemoprevention trials. *Clinical Breast Cancer,* 2001. 2(1): p. 33-36.

[80] Veronesi, U., et al., Tamoxifen for the prevention of breast cancer: late results of the Italian Randomised Tamoxifen Prevention Trial among women with hysterectomy. Journal of National Cancer Institute, *2007. 99(9): p. 727-737.*

[81] Cuzick, J., Aromatase inhibitors in prevention - data from the ATAC (arimidex, tamoxifen alone or in combination) trial and the design of IBIS-II (the second International Breast Cancer Intervention Study). *Recent Results Cancer Research,* 2003. 163: p. 96-103.

[82] Dowsett, M. and A. Howell, Breast cancer: Aromatase inhibitors take on tamoxifen. *Nature Medicine,* 2002. 8(1341-1344).

[83] Dowsett, M., et al., Retrospective analysis of time to recurrence in the ATAC Trial according to hormone receptor status: An hypothesis-generating study. *Journal of Clinical Oncology,* 2005. 23(30): p. 7512-7517.

[84] Dowsett, M., et al., Relationship between quantitative estrogen and progesterone receptor expression and human epidermal growth factor receptor 2 (HER-2) status with recurrence in the Arimidex, Tamoxifen, Alone or in combination trial. *Journal of Clinical Oncology,* 2008. 26(7): p. 1059-1065.

[85] Anderson, H., et al., Progesterone receptor levels are associated with time to aromatase inhibitor treatment failure in estrogen receptor-positive advanced breast cancer. *Cancer Research,* 2008. 69(Suppl.)(2): p. 6051.

[86] Shou, J., et al., Mechanisms of tamoxifen resistance: increased estrogen receptor-HER2/neu cross-talk in ER/HER2-positive breast cancer. *Journal of National Cancer Institute,* 2004. 96(12): p. 926-935.

[87] Mauriac, L., et al., Results of randomized phase II trial combining Iressa (Gefitinib) and Arimidex in women with advanced breast cancer (ABC). EORTC protocol 10021. *Cancer Research,* 2008. 69(Suppl.)(2): p. 6133.

[88] Gligorov, J., et al., Novel therapeutic strategies combining antihormonal and biological targeted therapies in breast cancer: Focus on clinical trials and perspectives. *Critical Reviews in Oncology/Haematology,* 2007. 64(2): p. 115-128.

[89] Marcom, P., et al., The combination of letrozole and trastuzumab as first or second line biological therapy produces durable responses in a subset of HER2 positive and ER positive advanced breast cancers. *Breast Cancer Research and Treatment,* 2007. 102: p. 43-49.

[90] Seidman, A., et al., Cardiac dysfunction in the Trastuzumab clinical trials experience. *Journal of Clinical Oncology,* 2002. 20(5): p. 1215-1221.

[91] Johnston, S., et al., Lapatinib combined with letrozole vs letrozole alone for front line postmenopausal hormone receptor positive (HR+) metastatic breast cancer (MBC): first results from the EGF30008 trial. *Cancer Research,* 2008. 69(Suppl.)(2): p. 46.

[92] Chan, C., et al., Molecular changes associated with the acquisition of oestrogen hypersensitivity in MCF-7 breast cancer cells on long-term oestrogen deprivation. *Journal of Steroid Biochemistry Molecular Biology,* 2002. 81(4-5): p. 333-341.

[93] Martin, L., et al., The anti-oestrogen ICI 182,780, but not tamoxifen, inhibits the growth of MCF-7 breast cancer cells refractory to long-term oestrogen deprivation through down-regulation of oestrogen receptor and IGF signalling. *Endocrine-Related Cancer,* 2005. 12(4): p. 1017-1036.

In: Aromatase Inhibitors: Types, Mode of Action and Indications ISBN: 978-1-60741-711-8
Editor: Jean R. Lamonte

Chapter VIII

Aromatase Inhibitors in Ovulation Induction. Myth or Reality?

Nikolaos P. Polyzos**, Spyridon Tzioras, Davide Mauri, Maria Tsappi, Antonis Valachis, Charalambos Dritsas, Kamposioras Konstantinos, Charalambos Panou and Lamprini Tsali

Panhellenic Association for Continual Medical research (PACMeR), Section of Obstetrics and Gynaecology, Greece

Abstract

Third generation aromatase inhibitors, drugs of choice for the treatment of hormone responsive breast cancer, have been introduced in clinical trials as potential agents for ovulation induction. Their efficacy has been tested against traditional ovulation induction regimens in women with anovulatory and other causes of infertility. However, while early data clearly supported the superiority of letrozole and anastrozole in ovulation induction, more recent publications' results were modest. In an attempt to clarify the discrepancies among published trials for aromatase inhibitors in ovulation induction we methodologically reviewed the available evidence and recorded quality characteristics. We separately assessed the level of evidence provided; sample size, statistical design, primary outcomes, secondary outcomes, side-effects and fetal outcomes in letrozole and anastrozole trials. Whereas an adequate number of Level I evidence was available, the statistical design and quality parameters were poorly reported. Primary outcomes (pregnancy rates) were not consistent among trials and data regarding secondary outcomes, side-effects and fetal outcomes were systematically omitted. Eventually, it is still unclear whether these drugs may represent a new treatment modality for ovulation

* Correspondence address: Nikolaos P. Polyzos M.D., Panhellenic Association for Continual Medical Research (PACMeR) Section of Obstetrics & Gynaecology, 28, Metamorfoseos St, 38333, Volos, Greece. Tel: +306936076146; Email: n.polyzos@gmail.com Conflict of interests: none to declare.

induction. Data is conflicting; trials are small in sample size and suffer from quality and statistical design biases. A large multicenter and properly designed randomized trial is required if a solid conclusion is to be made on the role of letrozole in ovulatory induction.

Introduction

Inhibition of aromatase enzyme, which catalyzes the conversion of androstenedione and testosterone into estrone and estradiol respectively, results in the reduced production of estrogens mainly in the ovaries and secondary and in other tissues such as the adipose and breast tissue, brain muscle and liver.

Aromatase inhibitors (AIs) have been initially introduced in the treatment of estrogen receptor (ER) positive breast cancer. Third generation AIs are highly potent and selective drugs compared to 1st and 2nd generation drugs. Recent evidence have demonstrated that third generation AIs, are associated with increased survival among patients with advanced breast cancer either used as first-line compared to tamoxifen or as second- and subsequent line compared to progestins (Mauri et al, 2006).

Mitwally and Casper initially conceived the idea of the utilization of 3rd generation AIs for ovulation induction. Early reports by these investigators (Mitwally and Casper, 2001 and Mitwally and Casper, 2002) supported that letrozole may result in acceptable ovulation and pregnancy rates in PCOS patients without affecting the endometrial thickness.

According to Mitwally and Casper, third generation AIs inhibit aromatization of androgens into estrogens without depleting the ERs as occurs with clomiphene citrate (CC) (Casper and Mitwally 2006). This decline in estrogen levels results in release of the hypothalamic-pituitary axis from the negative feedback and subsequently in an increased secretion of FSH. Thus, when administered in the early follicular phase (usually between 3rd and 7th cycle day) the rise in FSH levels results in growth of ovarian follicles. At the same time, the increase of intraovarian androgens results in promotion of follicular growth under the effect of FSH (Weil et al., 1998).

However the most advantageous effect of AIs is their short half-life (~45hours) compared to CC. Since they are rapidly cleared from the human body, their antiestrogenic effect ceases in the late follicular phase and as the dominant follicle grows the circulating estrogen levels rise. This results in a normal negative feedback centrally which leads to suppression of FSH levels and finally in atresia of small follicles and normo-ovulation. Moreover, since they are not associated with down-regulation of hypothalamic-pituitary estrogens receptors during the late follicular phase, adverse effects on peripheral organs such as endometrium and cervix are considerably less and might be associated with good pregnancy rates CC.

Since the initial reports by Mitwally and Casper, several controlled trials have been published to examine the role of letrozole and anastrozole in ovulation induction. Both 3rd generation AIs have been tested in women with anovulatory or unexplained infertility, alone or in combination with gonadotrophins and as a co-treatment in IVF/ICSI cycles.

The purpose of our systematic review is to summarize the available evidence for the use of 3rd generation AIs evaluate the quality of the information provided.

Search Strategy and Eligibility Criteria

We searched the Cochrane Central Trials Registry and PubMed without year and language restriction. The last search was updated on Dec 2008. We used the searching algorithm: (letrozole OR anastrazole OR aromatase OR aromatase inhibitors) AND (ovulation OR anovulation OR PCOS OR polycystic ovaries OR polycystic ovary syndrome OR infertility OR IVF OR ICSI OR in vitro fertilization). In addition, we tried to identify any previous systematic reviews in this field. We scrutinized the references of all eligible trials. Cross-searches were performed in Medline using the names of investigators who were lead authors in at least one eligible trial.

We considered eligible all controlled studies comparing 3rd generation aromatase inhibitors alone or in combination with other regimens versus any other ovarian stimulation regimens in infertile women for ovulation induction. Comparisons were eligible regardless of the doses, schedules and the cause of infertility. Trials were considered eligible whether women were planned to conceive with timed intercourse (TI), intrauterine insemination (IUI) or in vitro fertilization (IVF).

Data Extraction and Outcomes

From each eligible trial we recorded the following items: authors' names, journal and year of publication, the exact type dosage and schedules of regimens, the study sample size, number of patients per arm, pregnancy rates per arm, the type of infertility and whether patients were planned to conceive with TI, IUI of IVF/ICSI.

Main outcome measures were live birth rates and pregnancy rates. We recorded whether the authors provided live birth rates and clinical pregnancy rates (defined as the presence of intrauterine sac with embryonic pole demonstrating cardiac activity). We also recorded whether the author considered as clinical pregnancies, pregnancies defined as only the presence of intrauterine or rise in the levels of bHCG. We additionally recorded whether the authors provided data regarding secondary outcomes such as multiple pregnancies, miscarriages, ectopic pregnancies or cases of OHSS, data regarding adverse effects of medication and data for fetal outcomes.

Finally, we assessed level of evidence provided by recording whether the trial was randomized, prospective or retrospective. Whenever we came across a randomized controlled trial, we recorded study design items, including whether there was a description of the mode of randomization, allocation concealment and blinding.

Results

Characteristics of Included Trials

Overall thirty four studies compared an aromatase inhibitor versus other ovulation induction treatment in infertile women (Table 1). Of them 26 trials (76%) involved the comparison of letrozole versus other ovulation induction regimens, 4 (12%)the comparison of different schedules or doses of letrozole, 2(6%) the comparison of anastrozole versus other ovulation induction regimens and 2 (6%) the comparison of letrozole versus anastrozole.

Table 2 summarizes baseline characteristics of the trials included in our analysis. Among 34 trials, in 11(32%) trials patients were planned to conceive with planned or timed intercourse, in 17 (50%) with IUI in 2 trials (6%) either with TI or IUI, in 3 trials (9%) with IVF or ICSI, while in one trial (3%) the authors did not specify whether patients were planned to undergo TI or IUI.

The cause of infertility differed in the trials that were considered eligible. Ten of the trials (29%) included only patients with PCOS, 15 trials (44%) only patients with unexplained infertility whereas 9 trials (27%) included patients with more than two causes of infertility.

Outcome Measures

Primary and secondary outcomes are summarized in Table 1. Overall only 5 (15%) out of 34 eligible trials provided data regarding live births (Bayar et al.,2006a, Sohrabvand et al., 2006, Sipe et al., 2006, Gregoriou et al., 2007 and Bedaiwy et al., 2006). All of the trials, except one (Noriega-Portella et al., 2007) assessed clinical pregnancy rates. Nonetheless, only in 10 of the trials (30%), pregnancy was defined as the presence of intrauterine sac with an embryonic pole demonstrating cardiac activity. In 7 trials (21%) pregnancy was considered just the presence of intrauterine sac with or without a rise in bhCG levels, in 7 trials (21%) only a rise in the levels of bhCG values in two consecutive measurements, while in 8 trials (24%) the authors did not specify what they considered as a clinical pregnancy. In one trial (3%) the authors did not clarify whether they considered as a clinical pregnancy only the presence of an intrauterine sac or the presence of a sac with an embryonic pole with cardiac activity (Verpoest 2006), while in another (3%) the authors have not assessed pregnancy rates (Greisinger et al., 2008).

Secondary outcomes were not also consistently reported in eligible trials. In 24 of the eligible trials (71%) the authors reported the presence or not of multiple pregnancies among treated women, whereas in 20 (59%) of the trials the miscarriage rate was clearly stated. Ectopic pregnancies has been assessed only in 4 (12%) of the trials, while the incidence or not of OHSS in 7(21%). Finally the incidence or not of any side effects from medication with AIs has been reported in 6 (18%) trials, while outcomes for fetuses only in 3(9%) of the trials.

Table 1. Main and secondary outcomes

Author/year	Regimen	Total sample	Per Arm	Pregnancy (%)	Pregnancy definition	Multiple pregnancies	Miscarriages	Ectopic pregnancies	OHSS	Side effects	Fetus outcomes
Atay 2006	Letrozole 2.5	106	51	21.6	cardiac activity	yes	no	no	no	no	no
	Clomiphene 100		55	9.1							
Bayar 2006a	Letrozole 2.5	80	38	9.1	b-hCG	yes	yes	no	no	yes	no
	Clomiphene 100		36	7.4							
Sohrabvand 2006	Letrozole 2.5 mg + Metformin	59	29	19	cardiac activity	yes	yes	no	no	no	yes
	Clomiphene 100mg + Metformin		30	7							
Badawy 2007a	Letrozole 5	438	218	15.1	b-hCG	yes	yes	no	yes	no	no
	Clomiphene 100		220	17.9							
Begum 2008	Letrozole 7.5	64	32	40.6	not specified	yes	yes	no	no	yes	no
	Clomiphene 100		32	18.75							
Baruah 2008	Letrozole 2.5-5mg	50	25	19	not specified	yes	yes	no	no	yes	no
	Clomiphene 100mg		25	12.5							
Sipe 2006	Anastrazole 1mg +FSH	50	25	12	cardiac activity	yes	no	no	no	no	no
	Clomiphene 100 mg +FSH		25	20							
Badawy 2007b	Anastrozole 1	216	115	10.2	b-hCG	yes	yes	no	yes	no	no
	Clomiphene 100mg		101	7.9							

Table 1. (Continued)

Author/year	Regimen	Total sample	Per Arm	Pregnancy (%)	Pregnancy definition	Multiple pregnancies	Miscarriages	Ectopic pregnancies	OHSS	Side effects	Fetus ourcomes
Sammour 2001	Letrozole 2.5	49	25	16.7	not specified	no	no	no	no	no	no
	Clomiphene 100mg		24	5.6							
Al-Fozan 2004	Letrozole 7.5	154	74	11.5	not specified	yes	yes	yes	no	no	no
	Clomiphene 100		80	8.9							
Bayar 2006b	Letrozole 2.5	46	21	10	not specified	yes	yes	no	no	yes	no
	Clomiphene 100mg		25	12							
Fatemi 2006	Letrozole 2.5	15	7	2:07	b-hCG	no	no	no	no	no	no
	Clomiphene 100mg		8	3:08							
Badawy 2008a	Letrozole 5	412	205	19	b-hCG	yes	yes	no	yes	no	no
	Clomiphene 100		207	18.3							
Wu 2007	Anastrozole 1	33	14	2	not specified	no	no	no	no	no	no
	Clomiphene 100mg		19	0							
Barroso 2006	Letrozole 2.5+rFSH 75IU	41	21	23.8	only gestational sac	yes	yes	no	no	no	no
	Clomiphene 100mg+ rFSH 75IU		20	20							
Jee 2006	Letrozole 2.5 + HMG	93	66	18.2	only gestational sac	yes	yes	yes	no	no	no
	Clomiphene 100mg+ HMG		27	25.9							

Table 1. (Continued)

Author/year	Regimen	Total sample	Per Arm	Pregnancy (%)	Pregnancy definition	Multiple pregnancies	Miscarriages	Ectopic pregnancies	OHSS	Side effects	Fetus ourcomes
Sh Tehrani Nejad 2008	Letrozole 5	140	70	32.8	only gestational sac	yes	yes	no	no	no	No
	Clomiphene 100		70	14.3							
Baysoy 2006	Letrozole 5	80	40	18.42	cardiac activity	yes	no	no	yes	yes	No
	HMG 75-150IU		40	15.78							
Gregoriou 2007	Letrozole 5	50	25	8.9	cardiac activity	yes	no	yes	no	no	No
	rFSH 150IU/2day		25	14							
Quintero 2007	Letrozole 5	85	52	7	cardiac activity	yes	yes	no	no	no	No
	FSH		33	18							
Mitwally 2003	Letrozole2.5mg + FSH 50-150IU	110	36	22.2	cardiac activity	no	no	no	no	no	No
	Clomiphene100mg+ FSH 50-IU		18	11.1							
	FSH 50-150IU		56	21.4							
Healey 2003	Letrozole5mg +FSH or HMG150IU	205	NR	21.6	b-hCG	yes	yes	no	no	no	No
	FSH or HMG150IU		NR	20.9							
Bedaiwy 2006	Letrozole2.5-5mg + FSH50-150IU	432	308	14,4	cardiac activity	yes	yes	no	no	no	yes

Table 1. (Continued)

Author/year	Regimen	Total sample	Per Arm	Pregnancy (%)	Pregnancy definition	Multiple pregnancies	Miscarriages	Ectopic pregnancies	OHSS	Side effects	Fetus ourcomes
	FSH50-150IU		124	15.9							
Noriega Portella 2007	Letrozole2.5mg+rFSH	110	26	NR	not specified	yes	yes	yes	no	no	Yes
	Letrozole 5mg=+rFSH		48	NR							
	rFSH		36	NR							
Al-Fadhili 2006	Letrozole 2.5mg	72	34	5.9	only gestational sac	yes	no	no	yes	no	No
	Letrozole 5mg		38	26.3							
Badawy 2007c	Letrozole 2.5mg	179	58	4.8	only gestational sac	no	yes	no	no	no	No
	Letrozole 5mg		61	4.3							
	Letrozole 7.5mg		60	6.5							
Griesinger 2008	Anastrozole 1mg	32	11	not assessed	not assessed	no	no	no	no	no	No
	Anastrozole 5mg		10	not assessed							
	Clomiphene 50mg		11	not assessed							

Table 1. (Continued)

Mitwally 2005	Letrozole 20mg (single dose)	NR	NR	15	cardiac activity	no	no	no	no	no	No
	Letrozole 2.5mg (3 to7 days)		NR	18							
	Letrozole 20mgx1 +FSH (single dose)		NR	20							
	Letrozole 2.5mg+ FSH (3 to7 days)		NR	16.7							
Badawy 2008b	Letrozole5mg (1-5d)	218	110	12.4	only gestational sac	yes	yes	no	yes	no	No
	Letrozole 2.5mg(1-10d)		108	17.4							
Al Omari 2004	Anastrozole 1mg	40	18	9.7	not specified	no	no	no	no	no	No
	Letrozole 2.5mg		22	18.8							
Badawy 2008c	Anastrozole 1mg	220	109	15.1	b-hCG	yes	yes	no	yes	no	No
	Letrozole 2.5mg		111	12.2							
Goswami 2004	Letrozole 2.5mg +rFSH 75IU	38	13	23	cardiac activity	no	no	no	no	yes	No
	GnRHa+ rFSH 300-450		25	24							
Garcia Velasco 2005	Letrozole2.5mg + FSH/HMG 225/150	147	71	22.4	only gestational sac	yes	yes	no	no	no	No
	FSH/HMG 225/150		76	15.2							
Veproest 2006	Letrozole+2.5 mg +rhFSH 150IU/day	20	10	50	unclear	no	no	no	no	no	No
	rhFSH 150IU/day		10	20							

A considerable percentage of trials, 23 of the trials (67%) were randomized controlled trials (Table 2). Two trials (6%) were quasi-randomized, 6 (18%) prospective controlled and 3 (9%) retrospective trials.

Table 2. Characteristics of eligible trials

	Author/year	Journal	Study type	TI/IUI/IVF	Infetility types
1	Atay 2006	J Int Med Res	RCT	TI	PCOS
2	Bayar 2006	Fertil Steril	RCT	TI	PCOS
3	Sohrabvand 2006	Hum Reprod	RCT	TI	PCOS
4	Badawy 2007a	Fertil Steril	RCT	TI	PCOS
5	Begum 2008	Fertil Steril	RCT	TI	PCOS
6	Baruah 2008	Arch Gynecol	Quasi random	TI	PCOS
7	Sipe 2006	Fertil Steril	RCT	IUI	PCOS, UNXPL
8	Badawy 2007b	Fertil Steril	Prospe	TI	PCOS
9	Sammour 2001	Fertil Steril	RCT	IUI	UNXPL
10	Al-Fozan 2004	Fertil Steril	RCT	IUI	UNXPL
11	Bayar 2006	Fertil Steril	Quasi-random	IUI/TI	UNXPL
12	Fatemi 2006	RBO	RCT	IUI	UNXPL
13	Badawy 2008a	Fertil Steril	RCT	IUI	UNXPL
14	Wu 2007	Gynecol Endocrinol	RCT	IUI	UNXPL
15	Barroso 2006	Fertil Steril	RCT	IUI	UNXPL
16	Jee 2006	Fertil Steril	Prospe	IUI	UNXPL
17	Sh Tehrani Nejad 2008	J Assist Reprod Genet	RCT	IUI	UNXPL
18	Baysoy 2006	RBMonline	RCT	IUI	UXPL
19	Gregoriou 2007	Fertil Steril	RCT	IUI	UXPL
20	Quintero 2007	Fertil Steril	Retro	IUI	PCOS, UNXPL, MF
21	Mitwally 2003	Hum reprod	prospe	IUI	UNXPL
22	Healey 2003	Fertil Steril	Retro	IUI	PCOS, UNXPL, MF
23	Bedaiwy 2006	Hum reprod	prospe	IUI	PCOS, UNXPL, MF,ENDO, OTHER
24	Noriega Portella 2007	Fertil Steril	retro	IUI	PCOS, UNXPL, MF, ENDO,OTHER
25	Al-Fadhili 2006	Fertil Steril	RCT	IUI	UNXPL

26	Badawy 2007c	RBMonline	RCT	TI	UNXPL
27	Griesinger 2008	Fertil Steril	RCT	NR	UNXPL
28	Mitwally 2005	Fertil Steril	prospe	IUI/TI	not specified
29	Badawy 2008b	Fertil Steril	RCT	TI	PCOS
30	Al Omari 2004	Int J Obstet Gynecol	RCT	TI	PCOS
31	Badawy 2008c	Fertil Steril	RCT	TI	PCOS
32	Goswami 2004	Hum reprod	RCT	IVF	not specified
33	Garcia Velasco 2005	Fertil Steril	Prospe	IVF	UNXPL, MF,TF, combination
34	Veproest 2006	RBMonline	RCT	IVF/ICSI	UNXPL, MF,TF

RCT; randomized controlled trial, Prospe; prospective, retro; retrospective, NR; not reported, TI; timed intercourse, IUI; intrauterine insemination, IVF; in vitro fertilization, UNXPL; unexplained, MF; male factor, ENDO; endometriosis.

Table 3. Quality parameters of eligible randomized controlled trials

Author/year	Randomization mode	Allocation concealment	Blinding
Atay 2006	no	no	no
Bayar 2006	yes	yes	DB
Sohrabvand 2006	yes	yes	SB
Badawy 2007a	yes	yes	no
Begum 2008	no	no	no
Sipe 2006	yes	yes	SB
Sammour 2001	no	no	DB
Al-Fozan 2004	yes	no	no
Fatemi 2006	yes	no	no
Badawy 2008a	yes	no	no
Wu 2007	yes	no	no
Barroso 2006	yes	yes	DB
Sh Tehrani Nejad 2008	no	no	no
Baysoy 2006	yes	no	no
Gregoriou 2007	no	no	no
Al-Fadhili 2006	yes	no	no
Badawy 2007c	yes	no	no
Griesinger 2008	yes	yes	SB
Badawy 2008b	yes	no	no
Al Omari 2004	no	no	DB
Badawy 2008c	yes	no	no

Table 3. (Continued)

Author/year	Randomization mode	Allocation concealment	Blinding
Goswami 2004	yes	yes	SB
Veproest 2006	yes	no	no

SB; single blind, DB;double blind.

Among randomized controlled trials, a sufficient number of trials, 17 (74%), reported an appropriate mode of randomization. Nonetheless, only 7 trials (30%) described a method to ensure allocation concealment, 4 (17%) were double blinded, 3 (13%) single-blinded while the majority of the trials 16 (70%) were not blinded neither for patients or the investigators (Table 3).

Conclusion

Our systematic review summarizes the available evidence regarding the use of 3rd generation aromatase inhibitors in ovulation induction. Over the last five years, a significant amount of trials has been published comparing AIs alone or in combination with other agents versus traditional ovulation induction regimens in women with different causes of infertility. Despite the satisfactory rate of randomized trials available, the number of patients enrolled in most of the trials was considerably small to reach safe conclusions. Additionally, considerable shortcomings emerge regarding the methodological quality of the trials since allocation concealment and blinding were not reported in the majority of the randomized trials included. Suboptimal reporting of methodological quality in randomized trials has been related to overestimation of an experimental intervention (Shulz et al., 1995). Nonetheless, this is not consistent in all medical domains (Ioannidis and Lau, 1998).

A substantial limitation is the fact that live birth rates were reported in a very small number of trials. Live birth has the most relevance for patients; however this is not always feasible since it requires time and effort to collect data and patients do not often return to fertility clinic after they are discharged (Daya, 2003).

Moreover, the definition of pregnancy was not homogeneous in the trials included. Less than one third of the studies considered as a pregnancy the presence of intrauterine sac with an embryonic pole demonstrating cardiac activity. All the other trials considered as pregnancies either only the presence of a gestational sac or a rise in bhCG levels. However, reporting of biochemical pregnancy (detection of bHCG in the blood) creates confusion because of the possible carry-over effect from exogenous administrated hCG that could be detected in blood test (Daya, 2003).

Secondary outcomes were systematically omitted in trials included. Except from multiple pregnancies and miscarriages which were reported in more than 50% of eligible trials, the incidence of ectopic pregnancies and OHSS was assessed in less than one third of the trials. Moreover, medication-related side effects and outcomes regarding fetuses were consistently missing despite the fact that all the trials were investigating the use of experimental drugs. It is likely that in many of the trials that the authors did not report secondary outcomes, no events were available due to the limited number of patients included. Nonetheless, future trials should focus on adequate reporting of secondary outcomes and pregnancy-related adverse effects.

Overall current literature regarding AIs in ovulation induction is incoherent and feeble. The quality of the trials published in this field has been previously underlined in two systematic reviews for the use of aromatase inhibitors in women with unexplained and anovulatory infertility (Polyzos et al., 2008a and Polyzos et al., 2008b). Whereas all of the trials in our review were pioneering trials of an investigational drug and they provided insightful directions, quality and methodological issues were unclear. Indeed, in most of the trials the number of patients enrolled was considerably small and trials were underpowered to detect reasonable differences in terms of pregnancy rates. Statistical design was suboptimal, allocation concealment and blinding were poorly reported, while most of the trials were not registered. Moreover live births were rarely reported and definition of pregnancies was not homogeneous in the majority of the trials. Thus, taking into consideration all the above limitations it would be optimistic to reach a safe conclusion regarding the use of letrozole in ovulation induction. The need of a large multicenter, properly designed and adequately powered trial is mandatory (Polyzos et al., 2008c). Clinical investigators, journals editors, public health institutions and aromatase inhibitors providers should work towards this direction if a solid conclusion is to be made on the role of AIs in ovulation induction.

References

Al-Fadhli R, Sylvestre C, Buckett W, Tan SL, Tulandi T (2006) A randomized trial of superovulation with two different doses of letrozole *Fertil Steril* 85,161-164.

Al-Fozan H, Al-Khadouri M, Tan SL, Tulandi T (2004) A randomized trial of letrozole versus clomiphene citrate in women undergoing superovulation. *Fertil Steril* 82, 1561-1563.

Al-Omari WR, Sulaiman WR, Al-Hadithi N (2004) Comparison of two aromatase inhibitors in women with clomiphene-resistant polycystic ovary syndrome *Int J Gynaecol Obstet* 85,289-291.

Atay V, Cam C, Muhcu M, Cam M, Karateke A. (2006) Comparison of letrozole and clomiphene citrate in women with polycystic ovaries undergoing ovarian stimulation. *J Int Med Res* 34, 73-76.

Badawy A, Aal IA, Abulatta M (2007a) Clomiphene citrate or letrozole for ovulation induction in women with polycystic ovarian syndrome: a prospective randomized trial *Fertil Steril Jun* 18 doi:10.1016/j.fertnstert.2007.02.062.

Badawy A, Aal IA, Abulatta M (2007b) Clomiphene citrate or anastrozole for ovulation induction in women with polycystic ovary syndrome? *A prospective controlled trial Fertil Steril.* Dec 29 doi:10.1016/j.fertnstert.2007.08.034.

Badawy A, Metwally M, Fawzy M (2007c) *Randomized controlled trial of three doses of letrozole for ovulation induction in patients with unexplained infertility Reprod Biomed Online* 14, 559-562.

Badawy A, Elnashar, A ,Totongy, M (2008a) Clomiphene citrate or aromatase inhibitors for superovulation in women with unexplained infertility undergoing intrauterine insemination: a prospective randomized trial. *Fertil Steril* in press.

Badawy, A, Mosbah, A.,Tharwat, A., Eid, M. (2008b) Extended letrozole therapy for ovulation induction in clomiphene-resistant women with polycystic ovary syndrome: a novel protocol. *Fertil Steril* in press.

Badawy A, Mosbah A, Shady M (2008c) Anastrozole or letrozole for ovulation induction in clomiphene-resistant women with polycystic ovarian syndrome: a prospective randomized trial. *Fertil Steril* 89, 1209-1212.

Barroso G, Menocal G, Felix H, Rojas-Ruiz JC, Arslan M, Oehninger S (2006) Comparison of the efficacy of the aromatase inhibitor letrozole and clomiphene citrate as adjuvants to recombinant follicle-stimulating hormone in controlled ovarian hyperstimulation: a prospective, randomized, blinded clinical trial. *Fertil Steril* 86, 1428-1431.

Baruah J, Roy KK, Rahman SM, Kumar S, Sharma JB, Karmakar D (2008) Endometrial effects of letrozole and clomiphene citrate in women with polycystic ovary syndrome using spiral artery Doppler. Arch Gynecol Obstet. doi10.1007/s00404-008-0714-4

Bayar U, Basaran M, Kiran S, Coskun A, Gezer S.(2006a) Use of an aromatase inhibitor in patients with polycystic ovary syndrome: a prospective randomized trial. *Fertil Steril;* 86:1447-51.

Bayar U, Tanriverdi HA, Barut A, Ayoglu F, Ozcan O, Kaya E (2006b) Letrozole vs. clomiphene citrate in patients with ovulatory infertility. *Fertil Steri*l 85, 1045-1048.

Baysoy A, Serdaroglu H, Jamal H, Karatekeli E, Ozornek H, Attar E (2006) Letrozole versus human menopausal gonadotrophin in women undergoing intrauterine insemination *Reprod Biomed Online* 13, 208-212.

Bedaiwy MA, Forman R, Mousa NA, Al Inany HG, Casper RF (2006) Cost-effectiveness of aromatase inhibitor co-treatment for controlled ovarian stimulation. *Hum Reprod* 21, 2838-2844

Begum MR, Ferdous J, Begum A, Quadir E. (2008) Comparison of efficacy of aromatase inhibitor and clomiphene citrate in induction of ovulation in polycystic ovarian syndrome. *Fertil Steril.* doi:10.1016/j.fertnstert.2007.08.044.

Casper RF, Mitwally MF. (2006) Review: aromatase inhibitors for ovulation induction. *J Clin Endocrinol Metab* 91, 760-771.

Daya S (2003). Pitfalls in the design and analysis of efficacy trials in infertility. *Hum Reprod* 18(5), 1005-1009.

Fatemi HM, Kolibianakis E, Tournaye H, Camus M, Van Steirteghem AC, Devroey P (2003) Clomiphene citrate versus letrozole for ovarian stimulation: a pilot study. *Reprod Biomed* Online 7, 543-546.

Garcia-Velasco JA, Moreno L, Pacheco A, Guillén A, Duque L, Requena A, Pellicer A(2005) The aromatase inhibitor letrozole increases the concentration of intraovarian androgens and improves in vitro fertilization outcome in low responder patients: a pilot study *Fertil Steril* 84, 82-7.

Goswami SK, Das T, Chattopadhyay R, Sawhney V, Kumar J, Chaudhury K, Chakravarty BN, Kabir SN (2004). A randomized single-blind controlled trial of letrozole as a low-cost IVF protocol in women with poor ovarian response: a preliminary report *Hum Reprod* 19, 2031-2035.

Gregoriou O, Vlahos NF, Konidaris S, Papadias K, Botsis D, Creatsas GK (2007). Randomized controlled trial comparing superovulation with letrozole versus recombinant follicle-stimulating hormone combined to intrauterine insemination for couples with unexplained infertility who had failed clomiphene citrate stimulation and intrauterine insemination *Fertil Steril* Oct 23 doi:10.1016/j.fertnstert.2007.06.099.

Griesinger G, von Otte S, Schultze-Mosgau A, Diedrich K, Schröer A (2008) Follicular and endocrine response to anastrozole versus clomiphene citrate administered in follicular phase to normoovulatory women: a randomized comparison.*Fertil Steril* Apr 2. doi:10.1016/j.fertnstert.2008.01.091.

Healey S, Tan SL, Tulandi T, Biljan MM(2003) Effects of letrozole on superovulation with gonadotropins in women undergoing intrauterine insemination. *Fertil Steril* 80, 1325-1329.

Ioannidis JP, Lau J. Can quality of clinical trials and meta-analyses be quantified? *Lancet* 1998 352, 590–591.

Jee BC, Ku SY, Suh CS, Kim KC, Lee WD, Kim SH.(2006) Use of letrozole versus clomiphene citrate combined with gonadotropins in intrauterine insemination cycles: a pilot study. *Fertil Steril.* 85, 1774-1777.

Mauri D, Pavlidis N, Polyzos NP, Ioannidis JP (2006) Survival with aromatase inhibitors and inactivators versus standard hormonal therapy in advanced breast cancer: meta-analysis *J Natl Cancer Inst* 98, 1285-1291.

Mitwally MF, Casper RF.(2001) Use of an aromatase inhibitor for induction of ovulation in patients with an inadequate response to clomiphene citrate. *Fertil Steril.* 75, 305-309.

Mitwally MF, Casper RF (2002) Aromatase inhibition improves ovarian response to follicle-stimulating hormone in poor responders *Fertil Steril* 77, 776-780.

Mitwally MF, Casper RF.(2003) Aromatase inhibition reduces gonadotrophin dose required for controlled ovarian stimulation in women with unexplained infertility. *Hum Reprod* 18, 1588-1597.

Mitwally MF, Casper RF (2005) Single-dose administration of an aromatase inhibitor for ovarian stimulation *Fertil Steril* 3, 229-231.

Casper RF, Mitwally MF (2006) Review: aromatase inhibitors for ovulation induction. *J Clin Endocrinol Metab.* 91, 760-771.

Noriega-Portella L, Noriega-Hoces L, Delgado A, Rubio J, Gonzales-Castañeda C,Gonzales GF (2007) Effect of letrozole at 2.5 mg or 5.0 mg/day on ovarian stimulation withgonadotropins in women undergoing intrauterine insemination *Fertil Steril* Dec 14. doi:10.1016/j.fertnstert.2007.08.060.

Polyzos NP, Tsappi M, Mauri D, Atay V, Cortinovis I, Casazza G (2008a). Aromatase inhibitors for infertility in polycystic ovary syndrome. The beginning or the end of a new era? *Fertil Steril* 89(2),278-80.

Polyzos NP, Tzioras S, Mauri D, Tsappi M, Cortinovis I, Tsali L, Casazza G (2008b). Treatment of unexplained infertility with aromatase inhibitors or clomiphene citrate: a systematic review and meta-analysis *Obstet Gynecol Surv* 63, 472-479.

Polyzos NP, Mauri D, Tzioras S (2008c). Letrozole in ovulation induction: time to make decisions. Hum Reprod Update doi:10.1093/humupd/dmn050.

Quintero RB, Urban R, Lathi RB, Westphal LM, Dahan MH (2007) A comparison of letrozole to gonadotropins for ovulation induction, in subjects who failed to conceive with clomiphene citrate *Fertil Steril* 88, 879-885.

Sammour A, Biljan MM, Tan SL, Tulandi T.(2001) Prospective randomized trial comparing the effects of letrazole (LE) and clomiphene citrate (CC) on follicular development, endometrial thickness and pregnancy rate in patients undergoing super-ovulation prior to intrauterine insemination (IUI). *Fertil Steril* 76:S110.

Sh Tehrani Nejad E, Abediasl Z, Rashidi BH, Azimi Nekoo E, Shariat M, Amirchaghmaghi E.(2008) Comparison of the efficacy of the aromatase inhibitor letrozole and clomiphen citrate gonadotropins in controlled ovarian hyperstimulation: a prospective, simply randomized, clinical trial. *J Assist Reprod Genet.;* 25, 187-190.

Schulz KF, Chalmers I, Hayes RJ, Altman DG (1995). Empirical evidence of bias. Dimensions of methodological quality associated with 609 estimates of treatment effects in controlled trials. *JAMA* 273:, 408–412.

Sipe CS, Davis WA, Maifeld M, Van Voorhis BJ (2006) A prospective randomized trial comparing anastrozole and clomiphene citrate in an ovulation induction protocol using gonadotropins *Fertil Steril* 86, 1676-1681.

Sohrabvand F, Ansari Sh, Bagheri M. (2006) Efficacy of combined metformin-letrozole in comparison with metformin-clomiphene citrate in clomiphene-resistant infertile women with polycystic ovarian disease. *Hum Reprod* 21, 1432-1435.

Verpoest WM, Kolibianakis E, Papanikolaou E, Smitz J, Van Steirteghem A, Devroey P (2006) Aromatase inhibitors in ovarian stimulation for IVF/ICSI: a pilot study *Reprod Biomed Online* 13, 166-172.

Weil SJ, Vendola K, Zhou J, Adesanya OO, Wang J, Okafor J, Bondy CA. (1998) Androgen receptor gene expression in the primate ovary:cellular localization, regulation, and functional correlations. *J Clin Endocrinol Metab.* 83, 2479-2485.

Wu HH, Wang NM, Cheng ML, Hsieh JN (2007) A randomized comparison of ovulation induction and hormone profile between the aromatase inhibitor anastrozole and clomiphene citrate in women with infertility. *Gynecol Endocrinol* 23, 76-81.

In: Aromatase Inhibitors: Types, Mode of Action and Indications ISBN: 978-1-60741-711-8
Editor: Jean R. Lamonte

Chapter IX

The Use of Aromatase Inhibitors in Ovarian Cancer and Identification of Responsive Tumors

S. P. Langdon[*], *G. Walker, D. Faratian, C. Gourley, A. R. W. Williams, I. Um, J. F. Smyth and D. J. Harrison*
Edinburgh Breakthrough Breast Unit, Institute for Genetics and Molecular Medicine, University of Edinburgh, EH4 2XU, United Kingdom

Abstract

It has long been recognised that many ovarian cancers express estrogen receptors and their growth may be sensitive to anti-estrogen strategies. Aromatase is present in normal ovaries and in between 33 and 81% of ovarian cancers; its inhibition will result in a reduction of both circulating and local levels of estrogen. Recently, several aromatase inhibitors have been investigated in phase II clinical trials of recurrent ovarian cancer. Five trials of letrozole have now been completed in ovarian cancer with consistent and encouraging results. A significant percentage of patients derive some clinical benefit as indicated by change in serum levels of CA125, the surrogate biomarker of ovarian cancer tumor load. Between 10 and 36 % of patients have a CA125 response while a further 18 to 36% patients achieve stabilisation of CA125. The use of ovarian cancer cell line models has helped identify panels of estrogen-regulated genes whose expression can be linked with response. Several studies have now demonstrated the feasibility of identifying a sensitive patient subgroup by measurement of these biomarkers within primary cancers and their semi-quantitative expression can be linked with the surrogate marker of disease response and stabilisation after letrozole treatment. This review

[*] Correspondence: Dr Simon Langdon, Edinburgh Breakthrough Breast Unit and Division of Pathology, University of Edinburgh, Western General Hospital, Edinburgh EH4 2XU, United Kingdom. Tel: 44-131-537-1763 ; Fax: 44-131-537-3159 ; E-mail: simon.langdon@ed.ac.uk

summarises the clinical studies undertaken to date in ovarian cancer and describes the progress made towards predicting a cohort who will respond to treatment.

Introduction

Ovarian cancer is the most lethal gynecological malignancy in the Western World. In the USA, approximately 23,000 new cases of ovarian cancer are identified each year and over 15,000 women will die of their disease [1]. The majority of patients (70%) will have advanced disease at presentation and chemotherapy is required following optimal debulking surgery to delay tumor regrowth [2]. While platinum and taxane based drugs are the primary drugs used to treat systemic disease, resistance frequently develops and further treatments are generally necessary [2]. Among the options for treatment of relapsed disease, endocrine agents have been trialled and have shown activity in a subgroup of patients [3-6]. These agents are relatively non-toxic and offer a potential attractive alternative to chemotherapy. Endocrine therapy in the form of tamoxifen has been used to treat epithelial ovarian cancer since the mid-1960s [3-6] but it is only recently with the introduction of aromatase inhibitors that studies have focussed on identifying which patients might benefit from anti-estrogen manipulation [7-10]. With the increasing use of aromatase inhibitors in breast cancer, there has been interest in the application of these agents to ovarian cancer. This chapter will focus on the evidence that the growth of ovarian cancers may be sensitive to estrogen (and therefore manipulation by aromatase inhibitors) and review the studies undertaken to identify which tumors might be responsive to these agents.

Estrogen Responsiveness of Ovarian Cancer

Estrogen receptors (ERs) are present in the majority of ovarian cancers with the percentage incidence being dependent on the technique of measurement used and level of expression considered positive. A large number of studies have measured expression and these are reviewed in Slotman and Rao [11]. ERα (estrogen receptor alpha) expression in ovarian cancer can be highly variable both within and between different histological subtypes and this is illustrated in Figure 1 for a series of ovarian cancers collected within Edinburgh, Scotland. The cut-point shown indicates 20 fmol/mg protein which was the level set for treatment of breast cancers with endocrine therapy and the data suggest that many ovarian cancers have levels of ERα suggesting that signaling through the receptor does occur. The biochemical assays which were used widely for measuring ERα content in tumors such as breast cancer in the clinical setting have now been replaced by semi-quantitative immunohistochemical measurements allowing measurement on formalin-fixed tissue (Figure 2), and more recently by quantitative immunofluorescence technologies such as AQUA (HistoRx) (Figure 3).

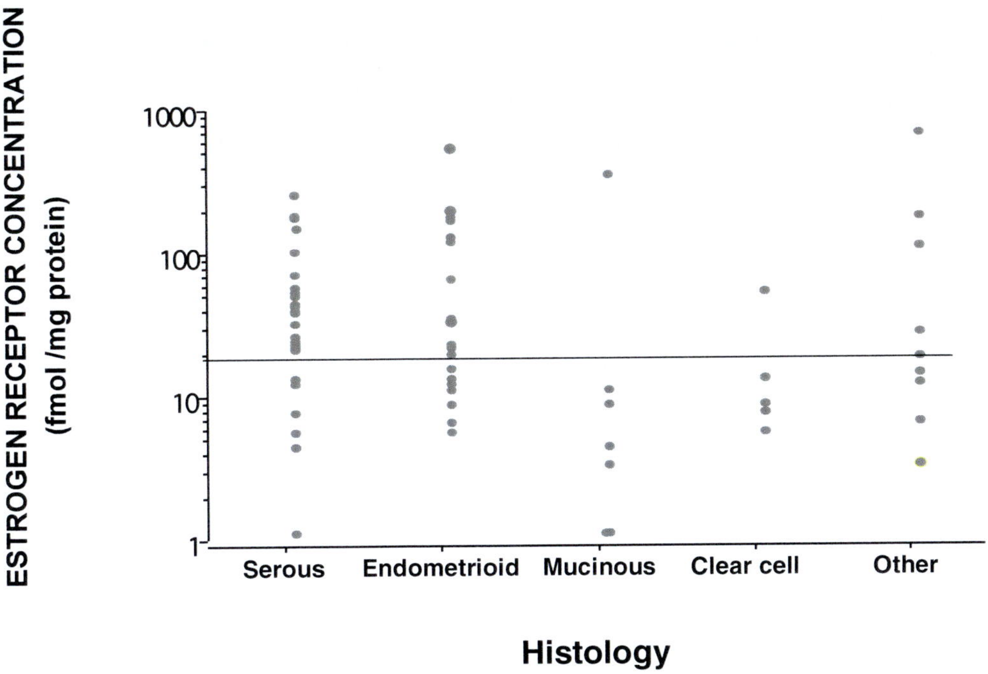

Figure 1. ERα expression in a series of ovarian cancers.
Each point represents a different tumor. Values measured by ER enzyme immunoassay. Different histological groups are compared.

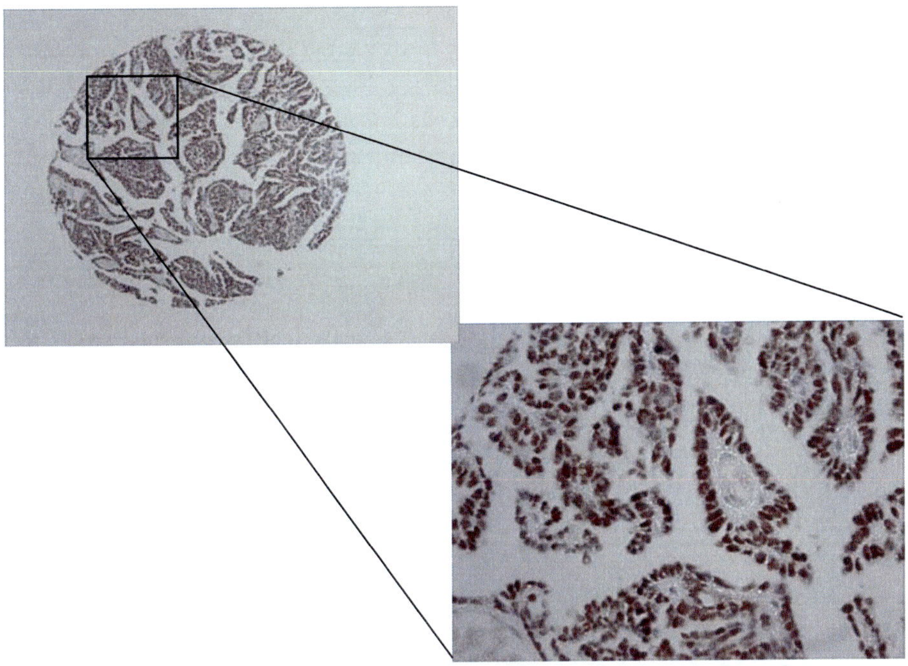

Figure 2. Immunohistochemical staining of an ovarian cancer showing expression of ERα.
A tissue array core is shown with brown staining indicating the presence of ERα expression in the nucleus of ovarian cancer cells.

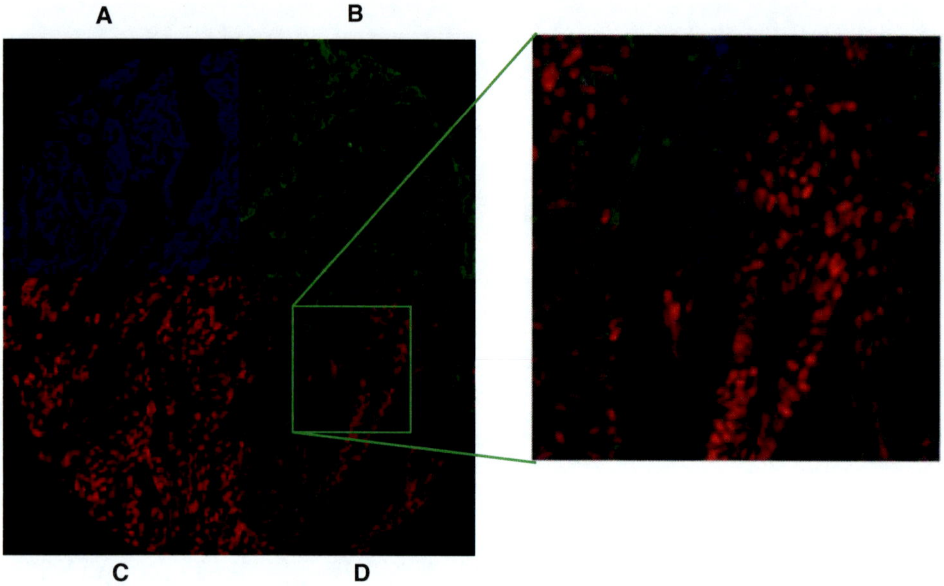

Figure 3. AQUA fluorescent image of an ovarian cancer TMA core showing expression of ERα. AQUA (Automated Quantitative Analysis) fluorescent analysis of ERα expression in a tissue microarray core, showing nuclear localisation of ERα (red) (C) and masking of tumor areas for quantitation by cytokeratin (green) (B) and nuclear staining by DAPI (blue) (A). The merged image is shown in D.

Direct evidence of estrogen sensitivity in ovarian cancer has been shown by the responses of ovarian cancer cells growing *in vitro*. Cell lines possessing high levels of ERα expression are predominantly growth responsive to estrogen and anti-estrogen manipulation while cell lines with low levels of expression are insensitive to these modulators [12-14] (Figure 4). Use of ERα and ERβ (estrogen receptor beta) targeted agonists suggested that both estrogen-regulated growth and gene expression act via ERα and not ERβ [15]. Indeed, while the growth promoting effects of estrogen are strongly linked to ERα, a tumor-suppressive role has been suggested for ERβ [16]. Expression of ERβ is high in normal ovary but ERα tends to be low. The ERα:ERβ ratio increases with progression to malignancy [16-18].

Epidemiological studies have also supported the notion that estrogen might be growth stimulatory. The "Million Women" study recently reported that estrogen in the form of hormone replacement therapy (HRT) caused an increased risk of ovarian cancer in postmenopausal women [19]. In that study, almost 950,000 women were followed for an average of 5.3 years for incidence and 6.9 years for death. Users of HRT had an increased incidence rate of 1.2 fold and increased death rate of 1.23 fold. Among the factors associated with increased risk are dose and length of exposure [19]. These numbers equate to an additional 1300 ovarian cancer cases and 1000 associated deaths since 1991 in the UK. This study is consistent with a number of other studies which have demonstrated an increased relative risk typically of about 1.5 [20,21].

Cell line	ER (fmol/mg)	Growth stimulation by E_2
PE01	130	+
PE04	209	+
PE06	81	+
PEA1	18	-
PEA2	8	-
PE014	5	-
PE023	8	-
PE016	8	-

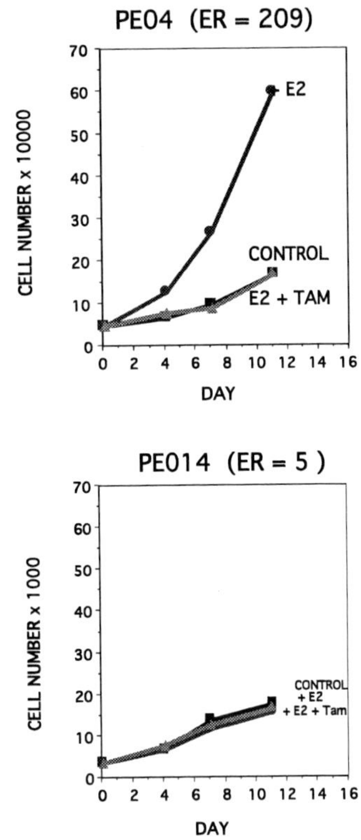

Figure 4. Model systems used to investigate estrogen regulation within ovarian cancer.

Clinical Use of Tamoxifen in Ovarian Cancer

Tamoxifen has been used widely to treat ovarian cancer with a 10-15% response rate, while other patients may derive some clinical benefit through disease stabilisation [3-5]. A Cochrane database analysis summarised 623 patients treated within 14 studies and reported response rates varying between 0 and 56% with 32% of patients showing stable disease for 4 weeks or longer [6]. The largest study, consisting of 105 patients suggested that the patients responding possessed a higher level of ERα expression but this did not achieve statistical significance [5]. These studies have demonstrated that a subgroup of women with ovarian cancer have disease that is tamoxifen-responsive. Extrapolation of clinical trial data from breast cancer would suggest that such a group might also benefit from the use of aromatase inhibitors.

Aromatase Expression in Ovarian Cancer

Expression of aromatase is frequently detected in ovarian cancer and it is feasible that local conversion of androgen to estrogen is important in the growth regulation of at least certain cancers [22]. Aromatase expression in ovarian cancer [22] has been demonstrated by the presence of mRNA [23,24], protein [23,25.26] and also by measurement of activity [25, 27,28]. Expression varies and the frequency of positive cancers ranges from 33 to 81% [22,23,25,26,27]. While the majority of studies have described no clear association between aromatase and steroid receptor expression [22,25,27,28], some studies have shown an inverse association between aromatase and ERα expression [24] and a positive association with the progesterone receptor have been reported [25,27,28]. High levels of estrogen can be produced within ovarian cancers and these can produce a variety of symptoms that resolve when the tumor is removed [29].

Further evidence that aromatase expression may be linked to an increased risk of ovarian cancer has been obtained from an analysis of genetic variation in the CYP19A1 gene that encodes P450 aromatase [30]. Genetic variation in two CYP19A1 single nucleotide polymorphisms has been shown to produce an increase in estrogen levels in postmenopausal women [31]. In turn, an increased risk of ovarian cancer was observed in Caucasian women who had these variants suggesting an association between high levels of expression of aromatase and ovarian cancer [30].

Preclinical studies investigating aromatase inhibitors in ovarian cancer cell lines have suggested that aromatase inhibitors can have direct actions on ovarian cancer cells implying that the aromatase activity within these cells, rather than in other distant tissues in the body, is sufficient for functionality [32]. It is feasible that ovarian cancers that respond to aromatase inhibitors may be dependent upon the aromatase produced within the tumor, in addition to dependency on estrogen regulation from other sources of aromatase activity such as adipose tissue.

Clinical Trials of Aromatase Inhibitors in Ovarian Cancer

The third-generation aromatase inhibitor, letrozole, has been the aromatase inhibitor most frequently studied to date in ovarian cancer [7, 8, 33, 34, 35]. In breast cancer trials, letrozole is orally active and inhibits aromatase activity by > 99% [36] and suppresses plasma levels of estrone (E_1), 17β-estradiol (E_2) and estrone sulphate (E_1S) in postmenopausal women by 95.2, 98.8 and 98.9% respectively [37]. Five trials have now been reported of this drug in ovarian cancer and results have been consistent between these studies [7, 8, 33, 34, 35] (Table 1).

Table 1 Phase II clinical trials of letrozole in ovarian cancer

Trial reference	Total	CA125 responders %	CA125 stable%
Bowman et al. [7]	54	10	26
Smyth et al.[8]	42	17	26
Papadimitriou et al.[33]	21	15	18
Kavanagh et al.[34]	13	31	39
Tchekmedyian et al [35]	14	36	36

The first phase II trial of letrozole in ovarian cancer was reported in 2002 by Bowman et al. [7]. That study demonstrated a 9.4% CA125 response rate and 26% CA125 stabilisation rate in a cohort of 54 patients. Measurement of the serum biomarker CA125 is now routine practice alongside use of the standard RECIST (Response Evaluation Criteria in Solid Tumors) criteria which measure radiological response [38,39]. A 20% disease stabilisation rate based on RECIST criteria was observed. While a number of biomarkers could be associated with CA125 response and stabilisation (see below), the most informative was ERα expression. Based on this study, an extension trial followed, with patients selected on increased ERα expression and this resulted in a 17 % CA125 response rate (7 of 42) with 26% of patients (11 of 42) exhibiting CA125 stabilisation [8]. In that study, there was a significant association between CA125 response and ERα expression level. A similar response rate of 15% and stabilisation rate of 18% was reported by Papadimitriou et al in a study of 21 patients [33]. Two smaller studies have provided further supportive data [34,35]. In one of these studies, 10 of 14 chemoresistant patients with recurrent disease had sustained stability or evidence of clinical improvement while on treatment with an average duration of 12 months [34]. In the other trial, 13 patients with tumors of low malignant potential were treated: 4 patients demonstrated a CA125 response (2 complete) (31%) and 5 demonstrated CA125 stabilisation (38%) [35].

There is still very limited data available for the other two widely used 3rd generation aromatase inhibitors, anastrozole and exemestane [22,40,41,42]. The single trial of exemestane reported to date recorded disease stabilisation in 36% of 22 patients but no responses [43]. Similarly, anastrozole has been reported to produce high rates of disease stabilisation - 42% of patients as reported by Del Carmen in a study of 53 patients with a single response [42]. In a study of combined anastrozole and the EGF receptor inhibitor gefitinib (Iressa), a stable disease rate of 61% in 23 patients (with a single complete response) was obtained [44].

Association of Response with Biomarkers

Since only a percentage of patients benefit from endocrine therapy, there has been interest in identifying this subgroup and features associated with these tumors. Identification of useful biomarkers could help improve the probability of targeting truly endocrine-sensitive tumors.

The studies described by Bowman et al [7] and Smyth et al [8] have demonstrated associations between CA125 response and/or stabilisation produced by letrozole and expression of ERα. Previous trials using tamoxifen had suggested a possible linkage between response and ERα expression but the trial size had always been insufficient to demonstrate a statistically significant association [3-5]. The study described by Bowman and colleagues described an association between increased probability of response and increased ERα expression [7]. This association, well known for the use of endocrine therapy in the treatment of breast cancer, had not been previously demonstrated for ovarian cancer. The immunoscore of 150 (of a maximum of 300) was selected for an extension trial [8] and within that latter study an increasing probability of CA125 response was again observed with increasing ERα content (subgroups of 150-199, 200-249, 250-300) with response rates of 0%, 12% and 33% response rates respectively [8]. This not only confirmed that increased ERα levels were associated with response but that the quality of response, ie CA125 reduction as opposed to CA125 stabilisation, also increased with ERα content [8]. Other markers which previously been demonstrated to be informative in breast cancer such as progesterone receptor have shown association with response [7]; similarly, EGF receptor and Her2 / erbB2 expression were also associated with outcome [7]. This suggests that potential combination strategies which target not only ERα but also pathways such as the EGF receptor might be of values and indeed the combination of anastrozole with the EGF receptor inhibitor has shown a high (61%) CA125 stabilisation rate [44].

Use of microarray analysis has identified a series of estrogen-regulated transcript changes in ERα-positive ovarian cells in culture [15]. Many of these genes have been demonstrated to be ERα and not ERβ-regulated [15]. A number of the associated proteins have now been tested in sections from primary tumors subsequently treated with letrozole in one of the phase II trials [7,9,10]. Biomarkers identified include the insulin-like growth factor binding proteins (IGFBPs) [9]. Three family members (IGFBP3, IGFBP4 and IGFBP5) have been shown to be estrogen-regulated in vitro while IGFBP1, IGFBP2 and IGFBP6 are unaffected [9]. IGFBP3 and IGFBP5 are down-regulated by estrogen while IGFBP4 is up-regulated [9]. These effects are reversed in the presence of tamoxifen. Evaluation of these proteins in primary ovarian cancers that were subsequently treated with letrozole indicated that IGBP3 and IGFBP5 were decreased, and IGFBP4 increased, in primary cancers that were responsive to letrozole relative to those that progressed on treatment, consistent with an estrogen-drive in responsive tumors [9]. Other markers up-regulated by estrogen in culture include trefoil-factor 1 (TFF1; pS2), TFF3, TNF receptor-associated protein 1 (TRAP1), topoisomerase II alpha (TOP2A) and ubiquitin-conjugating enzyme E2C (UBE2C), while plasminogen activator urokinase (PLAU), beta-IG-H3 (BIGH3) and vimentin (VIM) are down-regulated [10] (Figure 5). Consistent with these patterns, ovarian cancers expressing these up- and down-regulated genes tend to be letrozole-sensitive [10]. Prospective trials are now needed to confirm the value of these biomarkers as indicators of response.

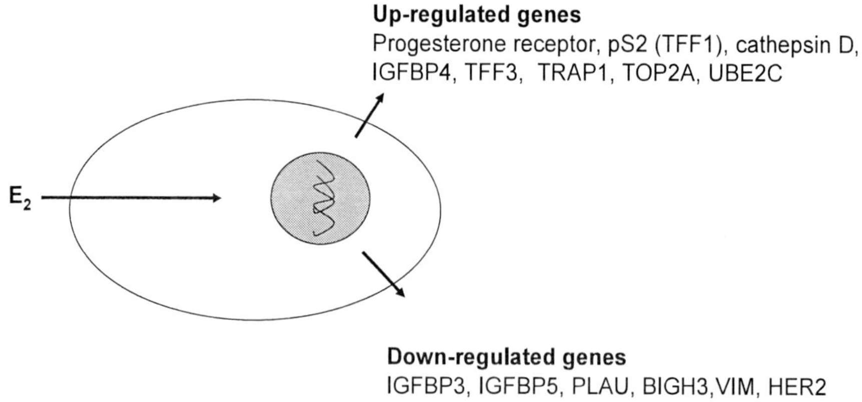

Figure 5. Genes modulated by E_2 which may represent biomarkers of letrozole response.

The presence of aromatase expression may be a relevant factor for anti-tumor response and increased expression has been reported to be present in letrozole-sensitive tumors [10]. This supports the idea that high expression of aromatase within cancers might be functional and the use of aromatase inhibitors might be more beneficial than anti-estrogens such as tamoxifen in particular patients.

Conclusion

It is clear from the trials undertaken with letrozole that aromatase inhibitors are effective in a subgroup of ovarian cancers. Current data suggests that these are most likely to be high expressing ERα-containing cancers, although other markers may also provide some indication of an estrogen-driven tumor. It is now be important to test letrozole in an adjuvant setting against low volume disease where its effects maybe more significant. In the setting of advanced disease where tamoxifen may elevate what is already a high thrombotic tendency, substitution of aromatase inhibitors with their more favourable side-effect profile may be appropriate.

Acknowledgments

We gratefully acknowledge support from the Charon Fund, Scottish Funding Council and Cancer Research UK which supported certain of the studies described in this review.

References

[1] Jemal A, Siegel R, Ward E, et al: Cancer statistics, 2007. *CA Cancer J Clin* 57:43-66, 2007.

[2] Ozols RF, Bookman MA, Connolly DC, et al. Focus on epithelial ovarian cancer. *Cancer Cell* 2004;5:19–24.

[3] J. Ahlgren, N. Ellison, R. Gottlieb, F. Laluna, J.J. Lokich, P.R. Sinclair, W. Ueno, G.L. Wampler, K.Y. Yeung, D. Alt and J.G. Fryer, Hormonal palliation of chemoresistant ovarian cancer: three consecutive phase II trials of the Mid-Atlantic Oncology Program, *J. Clin. Oncol.* 11 (1993), pp. 1957–1968.

[4] M. Markman, K. Webster, K. Zanotti, J. Rohl and J. Belinson, Use of tamoxifen in asymptomatic patients with recurrent small-volume ovarian cancer, *Gynecol. Oncol.* 93 (2004), pp. 390–393.

[5] K. Hatch, J. Beecham, J. Blessing and W. Creasman, Responsiveness of patients with advanced ovarian carcinoma to tamoxifen. A Gynecologic Oncology Group study of second-line therapy in 105 patients, *Cancer* 68 (1991), pp. 269–271.

[6] J.L. Perez-Gracia and E.M. Carrasco, Tamoxifen therapy for ovarian cancer in the adjuvant and advanced settings: systematic review of the literature and implications for future research, *Gynecol. Oncol.* 84 (2002), pp. 201–209.

[7] Bowman A, Gabra H, Langdon SP, et al. CA125 response is associated withestrogen receptor expression in a phase II trial of letrozole in ovarian cancer. Clin Cancer Res 2002; 8:2233–2239.

[8] Smyth JF, Gourley C, Walker G, et al. Antiestrogen therapy is active in selected ovarian cancer cases: the use of letrozole in estrogen receptorpositive patients. *Clin Cancer Res* 2007; 13:3617–3622.

[9] Walker G, MacLeod K, Williams ARW, et al. Insulin like growth factor binding proteins IGFBP3, IGFBP4 and IGFBP5 predict endocrine responsiveness in patients with ovarian cancer. *Clin Cancer Res* 2007; 13:1438–1444.

[10] Walker G, MacLeod K, Williams ARW, et al. Estrogen-regulated gene expression predicts response to endocrine therapy in patients with ovarian cancer. Gynecol Oncol 2007; 106:461–468.

[11] Slotman BJ, Rao BR. Ovarian cancer (review). Etiology, diagnosis, prognosis, surgery, radiotherapy, chemotherapy and endocrine therapy._Anticancer Res. 1988 May-Jun;8(3):417-34.

[12] S.P. Langdon, M.M. Hawkes, S.S. Lawrie, R.A. Hawkins, A.J. Tesdale and A.J. Crew *et al.*, Oestrogen receptor expression and the effects of oestrogen and tamoxifen on the growth of human ovarian carcinoma cell lines, *Br. J. Cancer* 62 (1990), pp. 213–216.

[13] S.P. Langdon, A.J. Crew, A.A. Ritchie, M. Muir, A. Wakeling and J.F. Smyth *et al.*, Growth inhibition of oestrogen receptor-positive human ovarian carcinoma by anti-oestrogens in vitro and in a xenograft model, *Eur. J. Cancer* 30A (1994), pp. 682–686

[14] S.P. Langdon, G.L. Hirst, E.P. Miller, A.R. Hawkins, A.L. Tesdale and J.F. Smyth *et al.*, The regulation of growth and protein expression by estrogen in vitro: a study of 8 human ovarian carcinoma cell lines, *J. Steroid Biochem. Mol. Biol.* 50 (1994), pp. 131–135.

[15] O'Donnell AJ, MacLeod KG, Burns DJ, et al. Estrogen receptor alpha mediates gene expression changes and growth response in ovarian cancer cells exposed to estrogen. *Endocr Rel Cancer* 2005; 12:851–866.

[16] Bardin A, Hoffmann P, Boulle N, Katsaros D, Vignon F, Pujol P & Lazennec G 2004 Involvement of estrogen receptor ß in ovarian carcinogenesis. *Cancer Research* 64 5861–5869.

[17] Rutherford T, Brown WD, Sapi E, Aschkenazi S, Munoz A & Mor G 2000 Absence of estrogen receptor-beta expression in metastatic ovarian cancer. *Obstetrics and Gynecology* 96 417–421.

[18] Pujol P, Rey JM, Nirde P, Roger P, Gastaldi M, Laffargue F, Rochefort H & Maudelonde T 1998 Differential expression of estrogen receptor-alpha and -beta messenger RNAs as a potential marker of ovarian carcinogenesis. *Cancer Research* 58 5367–5373.

[19] Beral V, Bull D, Green J, et al. Ovarian cancer and hormone replacement therapy in the Million Women Study. *Lancet* 2007; 369:1703–1710.

[20] C Rodriguez, AV Patel, EE Calle, EJ Jacob and MJ Thun, Estrogen replacement therapy and ovarian cancer mortality in a large prospective study of US women, *JAMA* 285 (2001), pp. 1460–1465.

[21] JV Lacey, LA Brinton and MF Leitzmann *et al.*, Menopausal hormone therapy and ovarian cancer risk in the National Institutes of Health-AARP Diet and Health Study Cohort, *J Natl Cancer Inst* 98 (2006), pp. 1397–1405.

[22] Li YF, Hu W, Fu SQ, et al. Aromatase inhibitors in ovarian cancer; is there a role? *Int J Gynecol Cancer* 2007

[23] Kaga K, Sasano H, Harada N et al. Aromatase in human common epithelial ovarian neoplasms. *Am J Pathol* 1996;149:45–51.

[24] Cunat S, Rabenoelina F, Daures JP et al. Aromatase expression in ovarian epithelial cancers. *J Steroid Biochem Mol Biol* 2005;93:15–24.

[25] Slotman BJ, Kuhnel R, Rao BR et al. Importance of steroid receptors and aromatase activity in the prognosis of ovarian cancer: high tumor progesterone receptor levels correlate with longer survival. *Gynecol Oncol* 1989;33:76–81.

[26] Noguchi T, Kitawaki J, Tamura T et al. Relationship between aromatase activity and steroid receptor levels in ovarian tumors from postmenopausal women. *J Steroid Biochem Mol Biol* 1993;44:657–60.

[27] Kitawaki J, Noguchi T, Yamamoto T et al. Immunohistochemical localisation of aromatase and its correlation with progesterone receptors in ovarian epithelial tumours. *Anticancer Res* 1996;16:91–7.

[28] Kuhnel R, Delemarre JF, Rao BR, Stolk JG. Correlation of aromatase activity and steroid receptors in human ovarian carcinoma. *Anticancer Res* 1986;6:889–92.

[29] Tokunaga H, Akahira J, Suzuki T, et al. Ovarian epithelial carcinoma with estrogen-producing stroma. *Pathol Int* 2007; 57:285–290.

[30] Goodman MT, Lurie G, Thompson PJ, McDuffie KE, Carney ME. Association of two common single-nucleotide polymorphisms in the CYP19A1 locus and ovarian cancer risk. *Endocr Relat Cancer.* 2008 Dec;15(4):1055-60.

[31] Haiman CA, Stram DO, Pike MC, Kolonel LN, Burtt NP, Altshuler D, Hirschhorn J & Henderson BE 2003 A comprehensive haplotype analysis of CYP19 and breast cancer risk: the Multiethnic Cohort. *Human Molecular Genetics* 12 2679–2692.

[32] Sasano H, Sato S, Ito K et al. Effects of aromatase inhibitors on the pathobiology of the human breast, endometrial and ovarian carcinoma. *Endocr Relat Cancer* 1999;6:197–204.

[33] Papadimitriou CA, Markaki S, Siapkaris J, et al. Hormonal therapy with letrozole for relapsed epithelial ovarian cancer. Long-term results of a phase II study. *Oncology* 2004; 66:112–117.

[34] Kavanagh JJ, Hu W, Fu S, et al. Antitumor activity of letrozole in patients with recurrent advanced low malignant potential or low-grade serous ovarian tumors. *J Clin Oncol* 2007; 25:5582 [ASCO Annu Meeting Proc Abs].

[35] Tchekmedyian NS, Liem AK, Quan ET, et al. Aromatase inhibitor therapy for estrogen receptor positive ovarian cancer. *J Clin Oncol* 2006; 24:15038 [ASCO Annu Meeting Proc Abs].

[36] Geisler J, Haynes B, Anker G, Dowsett M, Lønning PE.Influence of letrozole and anastrozole on total body aromatization and plasma estrogen levels in postmenopausal breast cancer patients evaluated in a randomized, cross-over study. *J Clin Oncol.* 2002 Feb 1;20(3):751-7.

[37] Geisler J, Helle H, Ekse D, Duong NK, Evans DB, Nordbø Y, Aas T, Lønning PE.Letrozole is superior to anastrozole in suppressing breast cancer tissue and plasma estrogen levels.*Clin Cancer Res.* 2008 Oct 1;14(19):6330-5.

[38] Rustin GJ, Bast RC Jr, Kelloff GJ, Barrett JC, Carter SK, Nisen PD, Sigman CC, Parkinson DR, Ruddon RW. Use of CA-125 in Clinical Trial Evaluation of New Therapeutic Drugs for Ovarian Cancer . *Clinical Cancer Research Vol.* 10, 3919-3926, June 1, 2004.

[39] Duffy MJ, Bonfrer JM, Kulpa J, Rustin GJ, Soletormos G, Torre GC, Tuxen MK, Zwirner M CA125 in ovarian cancer: European Group on Tumor Markers guidelines for clinical use *Int J Gynecol Cancer.* 2005 Sep-Oct;15(5):679-91.

[40] Krasner C. Aromatase inhibitors in gynaecologic cancer. *J Steroid Biochem Mol Biol* 2007; 106:76–80.

[41] Zheng H, Kavanagh JJ, Hu W, et al. Hormonal therapy in ovarian cancer. *Int J Gynecol Cancer* 2007; 17:325–338.

[42] del Carmen MG, Fuller AF, Matulonis U et al. Phase II trial of anastrozole in women with asymptomatic mullerian cancer. *Gynecol Oncol* 2003;91:596–602.

[43] Verma S, Alhayki M, Le T, et al. Phase II study of exemestane (E) in refractory ovarian cancer. *J Clin Oncol* 2006; 24:5026 [ASCO Annu Meeting Proc Abs].

[44] Krasner D, Debernado RL, Findley M; et al. Phase II trial of anastrazole in combination with gefitinib in women with asympomatic mullerian cancer. *J Clin Oncol* 2005; 23:5063 [ASCO Annu Meeting Proc Abst].

Index

A

AARP, 173
ABC, 17, 144, 146
abnormalities, 85, 92, 100
AC, 17, 52, 160
acetate, 17, 22, 34, 36, 49, 53, 104, 105, 118, 122, 142
acid, 107, 132, 143, 144, 145
acne, 9
ACTH, 9, 131
activation, x, 22, 24, 26, 29, 30, 33, 34, 50, 76, 77, 82, 83, 84, 86, 89
activators, 23
active site, 25, 26, 100
acute, 98, 110
Adams, 17
adaptation, 29, 36
adenine, 98
adenomyosis, ix, 75, 76, 77, 80, 81, 82, 83, 84, 86, 88, 90, 91, 92
adenosine, 99
adhesions, 96
adipose, vii, 1, 2, 5, 40, 43, 47, 48, 98, 99, 100, 148, 168
adipose tissue, vii, 1, 2, 5, 40, 43, 47, 48, 98, 99, 100, 168
adjunctive therapy, 102
adjustment, 8
administration, 5, 6, 16, 28, 38, 41, 45, 73, 87, 110, 161
adrenal gland, vii, 5, 24, 41
adrenal glands, 5, 24
adrenal insufficiency, 5
adult, 47
adverse event, 7, 43, 60, 63, 64, 66, 67, 68, 69, 70, 120, 123, 124, 126, 129, 131, 138
age, x, 3, 7, 8, 10, 23, 41, 47, 58, 60, 64, 66, 70, 95, 96, 99, 103
agent, viii, 3, 5, 6, 13, 23, 34, 39, 41, 43, 45, 104, 116, 141
agents, vii, viii, x, xi, 1, 5, 6, 11, 21, 39, 40, 76, 95, 97, 100, 105, 114, 137, 139, 147, 158, 164
aggressive behavior, 38
aggressiveness, 25, 77, 79
aging, 17
agonist, 23, 29, 92, 97, 101, 102, 103, 104, 105, 114, 123, 131, 139, 144
aid, 72
air, 2
AKT, 29, 33
aldosterone, 5, 9, 18, 41
algorithm, 149
ALP, 7
alpha, 2, 33, 34, 35, 36, 49, 50, 164, 170, 172
alternative, viii, 21, 24, 25, 32, 44, 45, 46, 47, 114, 120, 130, 137, 164
alternatives, 57
alters, 10
Amadori, 18
amelioration, 46
amenorrhea, 3, 83
Amsterdam, 91, 97, 102, 103, 104, 105
analgesic, 104
analgesics, 130
analog, 99
anatomy, 86
androgen, vii, 2, 5, 10, 15, 26, 41, 47, 48, 99, 102, 167

androgen receptors, vii, 2, 99
androgens, vii, viii, x, 1, 2, 5, 7, 12, 21, 22, 23, 25, 41, 44, 45, 47, 48, 95, 97, 98, 99, 100, 103, 148, 160
androstenedione, vii, x, 1, 12, 22, 25, 40, 41, 47, 48, 96, 98, 99, 115, 117, 148
angiogenesis, 24, 76, 85
angiogenic, x, 76, 85
anomalous, 81
antagonism, 25, 32, 37, 114
antagonist, 23, 31
antagonistic, 23, 34, 114
antagonists, 30, 110
anthracene, 47
antibody, 8, 31, 138
anticancer, 133
anticonvulsant, 40
anti-estrogenic, viii, 39, 40, 43, 44, 45, 47
anti-inflammatory agents, x, 95, 97
antithrombin, 14
antitumor, 7, 31
anti-tumor, 30
anti-tumor, 171
apoptosis, 24, 30, 76, 102, 108
application, 22, 89, 91, 105, 164
aromatases, 81
artery, 10, 160
arthralgia, 8, 43, 101, 126, 129, 138, 144
arthritis, 43, 126
ash, 13
Asian, 12, 45, 52
aspiration, 38
assessment, 17, 38, 102, 110, 133
assumptions, 61, 63, 66
asthenia, 41
asymptomatic, 171, 174
atherosclerosis, 130
Atlantic, 171
atresia, 148
availability, 32

B

babies, 46, 53
back, 41, 101, 102
back pain, 41, 101
Bangladesh, 39, 44
basic fibroblast growth factor, 30
bcl-2, 30
behavior, 38, 77, 78
Belgium, 44, 65
beneficial effect, 87
benefits, 4, 6, 12, 18, 23, 57, 63, 64, 66, 67, 68, 70, 92, 122
benign, x, 4, 95, 96, 141
bias, 25, 162
binding, vii, 1, 5, 23, 24, 25, 44, 88, 99, 100, 110, 111, 114, 115, 170, 172
binding globulin, 88
bioavailability, 16, 116
biological markers, 137
biomarker, xii, 70, 163, 169
biomarkers, xii, 33, 36, 163, 169, 170
biopsies, 34, 79, 140
biosynthesis, viii, x, 5, 16, 19, 39, 40, 41, 47, 48, 49, 91, 96, 97, 98, 99, 100, 106, 109, 116
biphosphonate, 64
birth, 3, 149, 158
birth rate, 149, 158
births, 150, 159
bisphosphonate treatment, 135
blastocyst, 3
bleeding, 4, 6, 7, 64, 81, 82, 83, 84, 85, 88, 93, 96, 101, 120, 123, 125, 126, 129
blocks, 5, 31, 45, 135
blood, viii, 8, 39, 42, 43, 53, 87, 158
blood clot, 8
blood flow, 43
bone age, 3
bone density, 8, 11, 102
bone loss, vii, 2, 7, 8, 17, 26, 90, 102, 123, 129, 132, 143, 144
bone mass, 7
bone remodeling, 8
bone resorption, 8, 23, 78
bowel, 96
brain, vii, 1, 2, 12, 24, 40, 98, 148
Brazil, 75
breast carcinoma, 3, 15, 37, 38, 48, 142
burn, 79

C

calcium, 102
cAMP, 50, 99, 100, 108
Canada, 44, 55, 60, 65, 70, 121, 129, 131, 136
cancer cells, 13, 28, 29, 30, 33, 34, 36, 37, 38, 50, 114, 115, 139, 146, 165, 166, 168, 172
cancer treatment, 11, 17, 36, 138
candidates, 42

carcinogenesis, 49, 172
carcinogenic, 42
carcinoma, 3, 15, 37, 48, 64, 136, 140, 142, 171, 172, 173
carcinomas, 3, 38
cardiac activity, 149, 150, 151, 153, 154, 155, 158
cardiomyopathy, 138
cardiovascular disease, 10
carpal tunnel syndrome, 129
catalytic activity, viii, 2
catechol, 50
Caucasian, 168
causal relationship, x, 76, 79
cDNA, 35
CEA, 56
cell, ix, xii, 19, 24, 25, 30, 33, 34, 75, 82, 89, 90, 98, 99, 108, 115, 135, 139, 163, 166, 168, 172
cell culture, 90, 108
cell cycle, 30
cell death, 33
cell line, xii, 24, 25, 29, 34, 135, 163, 166, 168, 172
cell lines, 24, 34, 135, 166, 168, 172
central nervous system, 5
cerebrovascular, 4, 6, 7, 122, 123, 126
cerebrovascular accident, 126
cervix, 43, 148
chemicals, vii, 2, 10
chemoprevention, xi, 4, 107, 113, 114, 135, 140, 145
chemopreventive agents, 11
chemoresistant, 169, 171
chemotherapy, 14, 25, 38, 73, 114, 128, 132, 133, 145, 164, 172
Chemotherapy, 145
Chi square, 81
chicken, 111
childhood, 3
Chi-square, 79, 83
cholesterol, 5, 9, 98, 99, 130
chromosome, 92
chronic pain, 104
circulation, 6, 41, 88
cis, 111
classes, 22
classical, 2, 24
classification, 25, 32
claudication, 130
cleavage, 46
clinical diagnosis, 81
clinical symptoms, 80
clinical trial, ix, xi, 4, 8, 11, 22, 28, 30, 32, 40, 41, 42, 46, 47, 55, 56, 60, 61, 64, 66, 71, 102, 105, 116, 118, 121, 127, 131, 134, 136, 146, 147, 160, 161, 162, 163, 167, 168
clinical trials, ix, xi, 4, 8, 11, 22, 30, 32, 40, 41, 46, 47, 55, 56, 60, 61, 64, 66, 71, 116, 118, 121, 127, 131, 136, 146, 147, 161, 163, 168
clinics, 110
co-activators, 23
coagulation, 46, 53
Cochrane, 145, 149, 167
coding, 3, 26
cofactors, viii, 21
cognitive performance, 107
cohort, xii, 44, 64, 66, 68, 88, 164, 169
colorectal cancer, 72, 143
combination therapy, 137, 140
combined oral contraceptives, x, 76
co-morbidities, 8, 12
competition, 100
complexity, 26
complications, 101, 128, 129
components, 10
compounds, viii, 21, 22, 25, 26, 30, 101
concealment, 149, 157, 158, 159
concentration, 6, 29, 43, 53, 99, 160
conception, 43
confusion, 158
consensus, 50
contraceptives, x, 76, 82, 83, 85, 90, 91, 92, 105
control, 32, 34, 44, 60, 99, 130
control group, 44
controlled research, ix, 40, 47
controlled studies, 149
controlled trials, 7, 119, 126, 135, 136, 137, 148, 150, 157, 158, 162
conversion, vii, x, 1, 5, 22, 23, 25, 40, 41, 45, 47, 48, 88, 89, 96, 98, 99, 100, 103, 115, 132, 148, 167
coronary artery disease, 10
coronary heart disease, 10
correlation, 52, 173
correlations, 162
corticosterone, 5
cortisol, 5, 9, 18, 41
cost effectiveness, 72
cost saving, 69

cost-effective, 56, 60, 61, 63, 65, 66, 67, 69, 70, 71
costs, ix, 55, 56, 60, 63, 64, 65, 66, 67, 69, 70, 71
couples, 52, 161
coupling, 50
covalent, 5
covalent bond, 5
CREB, 99, 108
critical analysis, 110
crossing over, 125
culture, 29, 53, 170
culture conditions, 29
curing, 77
cycles, 43, 45, 52, 53, 85, 86, 93, 148, 161
cyclic AMP, 108
cyclin D1, 50
cyclooxygenase, x, 76, 78, 92, 110
cyclooxygenase-2, 92, 110
cyst, 92
cysts, 4, 5, 99, 103, 104
cytochrome, 5, 9, 33, 38, 40, 41, 48, 91, 92, 98, 107, 109, 115
cytochrome p450, 40, 91, 92
cytokine, 82
cytokines, 31, 100
cytoplasm, 24, 82

D

database, 167
de novo, 99
death, 3, 4, 12, 19, 57, 64, 127, 128, 166
death rate, 4, 127, 166
deaths, 10, 166
debulking surgery, 164
decision making, 56, 63, 72
decisions, 63, 68, 71, 136, 162
deep venous thrombosis, 4
defects, 87
deficiency, 3, 13, 17
definition, 151, 152, 153, 154, 155, 158, 159
degradation, 31
dehydrogenase, 99, 111
density, vii, 2, 7, 8, 11, 17, 33, 36, 43, 102, 116, 123, 132, 135, 143, 145
deprivation, viii, 3, 22, 23, 29, 30, 32, 33, 36, 37, 39, 40, 123, 139, 146
derivatives, 22
detection, 79, 82, 84, 86, 114, 158
dexamethasone, 81, 91, 141
diarrhea, 101
differentiation, 100
dimerization, 31
direct action, 168
discomfort, 129
discrimination, 34
disease free survival, 32, 58, 125, 127, 128, 129, 132, 134, 136
disease progression, 31, 42, 120, 139
disease rate, 169
disease-free survival, 4, 42, 62, 101, 122, 126
dislocation, 89
divergence, 59
DNA, 24, 82, 115
doctors, 4
dominance, 70, 82
dominant strategy, 69
Doppler, 160
dosage, 149
dosing, 6
double-blind trial, 36, 43, 142
down-regulated genes, 170
down-regulation, 146, 148
drugs, vii, viii, xi, 2, 6, 8, 11, 25, 31, 39, 40, 60, 64, 78, 103, 104, 147, 148, 158, 164
duration, 56, 60, 62, 67, 118, 125, 127, 132, 169
dysmenorrhea, x, 80, 81, 82, 83, 89, 95, 96
dyspareunia, x, 95, 96, 129
dyspnea, 101
dysregulation, 93

E

egg, 53
elderly, 114, 134, 140, 141, 144, 145
electron, 5
emboli, 4
embryo, 43, 46, 53, 91, 93
endocrine, vii, viii, xi, 2, 5, 12, 15, 16, 18, 19, 21, 22, 23, 27, 28, 30, 31, 32, 34, 35, 36, 37, 42, 71, 72, 73, 106, 107, 110, 113, 114, 115, 119, 120, 121, 122, 123, 127, 129, 131, 132, 133, 134, 135, 137, 138, 139, 140, 141, 143, 144, 145, 161, 164, 169, 172
endocrine resistance, viii, 21, 28, 31, 35, 37, 138
endocrinological, 36, 45
endocrinology, 2, 16, 91, 141
endometrial cancer, 4, 6, 7, 8, 14, 89, 122, 123, 129
endometrial carcinoma, 64

endometrial glands, 80, 81, 82
endometrial hyperplasia, 23, 126
endometrial polyps, ix, 75, 76, 77, 80, 84, 86, 88, 93
endometriosis, vii, ix, x, xi, 42, 43, 51, 75, 76, 77, 78, 79, 80, 81, 84, 86, 87, 89, 90, 91, 92, 93, 95, 96, 97, 99, 100, 102, 103, 104, 105, 106, 107, 108, 109, 110, 111, 157
endometrium, ix, x, 43, 44, 45, 46, 47, 49, 53, 75, 76, 77, 78, 79, 80, 81, 82, 83, 84, 85, 86, 87, 88, 89, 90, 91, 92, 96, 106, 107, 108, 109, 111, 123, 148
endoplasmic reticulum, 98
enlargement, 82
environment, 45, 53
enzymatic, 2, 5, 77, 98, 107
enzyme immunoassay, 165
enzyme induction, 36
enzymes, x, 5, 9, 25, 40, 41, 76, 89, 92, 98, 99, 100
epidemiology, 19
epidermal growth factor, 14, 24, 29, 30, 34, 50, 146
epidermal growth factor receptor, 14, 24, 29, 34, 50, 146
epithelia, 109
epithelial cell, 23, 37, 102
epithelial cells, 23, 37, 102
epithelial ovarian cancer, 164, 171, 173
epithelial tumours, 173
epithelium, 81, 88
equality, 46
Erk, 34
estradiol, x, 14, 29, 31, 36, 37, 40, 41, 45, 48, 49, 50, 53, 83, 88, 95, 96, 98, 99, 100, 102, 111, 139, 148, 168
estrogen, vii, viii, ix, x, xi, 1, 2, 3, 4, 5, 6, 7, 8, 10, 11, 12, 13, 15, 16, 17, 19, 21, 22, 23, 24, 26, 28, 29, 30, 31, 33, 34, 35, 36, 37, 39, 40, 41, 42, 43, 44, 45, 46, 47, 48, 49, 50, 56, 72, 75, 76, 78, 79, 81, 82, 84, 85, 89, 90, 91, 92, 95, 96, 97, 98, 99, 100, 101, 102, 106, 107, 109, 110, 141, 142, 146, 148, 163, 164, 166, 167, 168, 170, 171, 172, 173, 174
estrogen receptor modulator, 34
estrogen receptors, viii, xi, 3, 21, 35, 37, 44, 45, 50, 90, 163
Estrogen replacement therapy, 173
estrogens, vii, viii, x, 1, 2, 3, 5, 12, 13, 15, 21, 22, 39, 40, 41, 42, 43, 45, 75, 77, 78, 80, 82, 84, 85, 86, 88, 89, 90, 91, 98, 99, 100, 102, 103, 108, 148, 171
etiology, 13
evolution, ix, 55, 71
excision, 28, 32
exposure, 25, 28, 39, 42, 75, 76, 81, 86, 89, 166
extracellular matrix, 78
extrapolation, 60
eye, 14

F

failure, 28, 36, 78, 85, 102, 108, 111, 119, 121, 130, 133, 137, 138, 142, 146
family, 110, 115, 170
family members, 170
family planning, 110
fasting, 9, 16
fat, viii, 7, 24, 39, 41, 115
fatigue, 101
FDA, 8, 42
feedback, ix, 5, 39, 41, 44, 45, 81, 103, 130, 148
females, 3
fertility, 3, 44, 81, 85, 86, 87, 158
fertilization, 46, 53, 85, 91, 93, 149, 157, 160
fetal, xi, 3, 147, 149
fetuses, 150, 158
fibrinogen, 14
fibroblast, 30
fibroblast growth factor, 30
fibroblasts, 2, 47, 85, 98
fibroids, 43
fine needle aspiration, 38
first generation, 23, 116
flavonoid, 10
flavonoids, 10
flow, 43, 78
focusing, 71
follicle, 44, 51, 52, 53, 99, 130, 148, 160, 161
follicle stimulating hormone, 52
follicles, 23, 40, 45, 46, 49, 104, 148
follicle-stimulating hormone, 51, 52, 53, 99, 130, 160, 161
follicular, ix, 39, 44, 45, 52, 130, 148, 161, 162
food, 10, 38
fracture, 7, 8, 33, 64, 101, 123
fractures, 17, 64, 123, 125, 126, 129, 135
FSH, 3, 41, 44, 45, 52, 99, 101, 104, 130, 131, 148, 151, 153, 155

G

G protein, 34
gastrointestinal, 10
gene, x, 3, 12, 13, 24, 25, 26, 33, 34, 36, 37, 38, 40, 50, 70, 76, 77, 78, 82, 83, 86, 98, 99, 108, 109, 162, 166, 168, 172
gene expression, 25, 34, 37, 38, 50, 162, 166, 172
gene promoter, 26, 76
generation, vii, viii, xi, 1, 5, 6, 7, 11, 18, 21, 22, 23, 27, 32, 39, 40, 41, 42, 45, 56, 71, 96, 100, 101, 104, 107, 109, 113, 115, 116, 117, 118, 119, 122, 126, 130, 134, 135, 139, 140, 141, 147, 148, 149, 158, 168, 169
genes, xii, 3, 13, 24, 28, 35, 50, 82, 89, 99, 108, 163, 170
genome, 50
genomic, viii, 2, 12, 21, 24, 29, 33, 35
genomics, 36
Germany, 44
gland, 47
globulin, 88
glucose, 3
glycoprotein, 138
gold, viii, 21, 22, 114, 115, 122
gold standard, viii, 21, 22, 114, 115, 122
gonadotropin, ix, x, 5, 39, 40, 44, 46, 47, 53, 95, 97, 104, 105, 108
gonadotropin secretion, ix, 5, 39, 44
gonadotropin-releasing hormone (GnRH), 97
government, iv
G-protein, 29, 99
granulosa cells, 2, 12, 98, 103, 115
grapes, 10
Greece, 44, 45, 147
groups, 42, 120, 123, 125, 127, 134, 137, 165
growth, viii, x, xi, 3, 10, 14, 21, 24, 28, 29, 30, 31, 32, 34, 37, 41, 42, 44, 45, 50, 52, 81, 84, 85, 90, 91, 95, 96, 97, 99, 107, 137, 139, 146, 148, 163, 164, 166, 167, 170, 172
growth factor, viii, 14, 21, 24, 28, 29, 30, 31, 34, 37, 50, 52, 90, 138, 146, 170, 172
growth factors, 29, 30, 31
growth rate, 85
growth spurt, 3
guidelines, 8, 27, 28, 60, 105, 174
gynecologists, 105

H

hair follicle, 2
half-life, 43, 148
haplotype, 173
haplotype analysis, 173
harm, 4
harmful effects, 43
hazards, 65
HDL, 9
head and neck cancer, 10
headache, 41, 101
health, vii, 2, 7, 8, 17, 56, 62, 64, 65, 70, 71, 92, 109, 143
health care, 56, 65, 70, 71
health care system, 70
heart, 10, 130
heat, 24
heat shock protein, 24
hemostasis, 53
HER2, 29, 30, 138, 146
Herceptin, 32
heterogeneity, 58, 60
heterogeneous, 60
high risk, 8, 11, 66, 67, 114, 135, 140
high-risk, 11, 60
hip, 17, 123
hip fracture, 123
hip fractures, 123
histological, 164, 165
histopathology, 92
homolog, 34
horizon, 62, 63, 65, 66, 69
hormonal therapy, ix, 14, 55, 56, 57, 59, 64, 65, 66, 73, 101, 161
hormone, vii, viii, x, xi, 1, 5, 6, 8, 11, 17, 21, 22, 27, 28, 30, 32, 33, 35, 36, 38, 41, 42, 49, 51, 52, 53, 57, 58, 61, 69, 71, 72, 73, 88, 92, 93, 95, 97, 99, 101, 110, 113, 114, 116, 119, 129, 130, 131, 132, 135, 137, 138, 139, 140, 143, 144, 145, 146, 147, 160, 161, 162, 166, 172, 173
hormones, 13, 17, 31, 49, 50, 87, 89
hospital, 45, 106
hospitalization, 96
HPV, 16
human, 3, 12, 13, 14, 24, 25, 33, 34, 35, 36, 37, 38, 40, 47, 48, 53, 81, 86, 89, 91, 92, 98, 107, 108, 109, 110, 138, 139, 146, 148, 160, 172, 173

human chorionic gonadotropin, 53
humans, 3
hydrocortisone, 15
hydrogen, 47
hydroxyl, 42
hydroxylation, 2, 5, 98, 100
hyperandrogenism, 3
hypercholesterolemia, 8, 9
hyperestrogenism, 81
hyperplasia, 23, 91, 126, 130
hypersensitivity, 29, 33, 36, 146
hypertension, 12, 130
hypertensive, 130
hypertrichosis, 9
hypothalamic, ix, 39, 44, 148
hypothalamic-pituitary axis, ix, 39, 44, 148
hypothalamus, 2, 41, 45, 103
hypothesis, ix, 26, 28, 44, 75, 77, 78, 81, 84, 86, 87, 88, 89, 146
hysterectomy, 4, 96, 146

I

IBIS, 107, 135, 145, 146
ICSI, 148, 149, 150, 157, 162
identification, vii, 1, 5
IES, 9, 10, 33, 59, 62, 64, 65, 121, 127, 128, 131, 136
IGF, 29, 30, 35, 146
immune response, 90
immune system, 91
immunofluorescence, 164
immunohistochemical, 79, 82, 88, 164
immunohistochemistry, 83
immunological, 78
implantation failures, 86
implants, viii, x, 39, 43, 95, 96, 97, 99
in situ, 100, 102
in vitro, 12, 24, 29, 31, 33, 34, 47, 48, 53, 85, 90, 93, 115, 141, 149, 157, 160, 166, 170, 172
in vitro fertilization, 53, 85, 93, 149, 157, 160
in vitro fertilization (IVF), 149
in vivo, 10, 16, 25, 34, 47, 99, 141
inactivation, 82, 107
inactive, 101
incidence, 4, 6, 7, 8, 10, 33, 43, 44, 60, 70, 88, 123, 125, 136, 150, 158, 164, 166
inclusion, 62
India, 44
indication, 5, 43, 90, 171
indicators, 170
indomethacin, 81, 91
inducer, 25
induction, xi, 36, 43, 44, 45, 50, 51, 52, 53, 54, 99, 147, 148, 149, 150, 158, 159, 160, 161, 162
infarction, 130
infertile, 45, 51, 149, 150, 162
infertility, ix, xi, 40, 43, 44, 46, 47, 51, 52, 86, 87, 89, 105, 108, 109, 110, 147, 148, 149, 150, 158, 159, 160, 161, 162
inflammation, ix, 75, 76, 78, 79, 81, 82, 84, 85, 88, 89, 96
inflammatory, ix, x, 75, 76, 77, 81, 82, 89, 90, 95, 97, 104
inflammatory mediators, ix, 75, 76, 77, 81
inflammatory response, 90
inflation, 71
inhibitory, 5, 82, 87, 116
inhibitory effect, 5, 87
initiation, 30, 32, 42, 49, 52, 77, 87
injections, 47, 104
insemination, 44, 52, 53, 149, 157, 160, 161, 162
institutions, 159
insulin, 3, 24, 29, 30, 52, 170
Insulin like growth factor, 172
insulin-like growth factor, 24, 29, 52, 170
integration, 56, 107
integrin, 108
interaction, 2, 37
interactions, 24, 26, 32, 52
interference, 22
interleukin-1, 110
intervention, 57, 60, 158
intramuscular, 5, 15, 116
intravenous, 123
intrinsic, 7
invasive, 126, 134, 135
Investigations, vii
ipsilateral, 57, 62
Ireland, 18
irritation, 4
ischaemic heart disease, 130
ischemic, 4, 123
isomers, 43
ITA, 9, 59, 72, 121, 127, 128, 143
IVF, 86, 87, 93, 148, 149, 150, 156, 157, 161, 162

J

JAMA, 162, 173
Japan, 1, 95, 105
joint pain, 130
Jordan, 22, 28, 35
judge, 57
jurisdiction, 65
jurisdictions, 56, 60

K

kappa, ix, 75, 76, 82, 83, 89
kappa B, 82
karyotype, 92
kidneys, 8
kinase, 24, 29, 30, 31, 33, 50, 115, 138
kinases, 24, 29
King, 16, 24, 34
knockout, 3
Korea, 44

L

laboratory studies, 135
language, 149
laparoscopic, 90, 104, 107
laparoscopy, 79, 86, 90, 104
large-scale, ix, 11, 40, 47
laser, 90
leiomyoma, 42, 92
Lesion, 79
lesions, x, 77, 78, 79, 80, 81, 82, 96, 97
lethargy, 116
Leydig cells, 2, 98
libido, 129
life expectancy, 70
life quality, 105
life span, 44
lifestyle, 12
life-threatening, 4, 7
lifetime, viii, 39, 42, 60
ligand, 8, 24
ligands, 26
lignans, vii, 2, 10
likelihood, 4, 62
limitation, 158
limitations, 4, 60, 68, 159
linear, 3
linkage, 169
lipid, 8, 9, 17, 18, 43, 116
lipid metabolism, 9, 43
lipid profile, 8, 9, 17, 116
lipids, 8, 17, 18, 35, 36
lipoprotein, 9
liver, viii, 2, 7, 39, 40, 41, 47, 48, 100, 115, 148
localised, 139
localization, 52, 162
locus, 173
London, 100
losses, 63
low risk, 66, 67
low-density, 9
low-density lipoprotein, 9
lubricants, 129
lumbar, 8
lumbar spine, 8
lung, 10
lung cancer, 10
luteinizing hormone, 101, 114, 144
lymph node, 13, 14, 143

M

Mackintosh, 110
macrophage, 78
macrophages, ix, 75, 78, 90
magnetic, iv, 18
males, 12
malignancy, 164, 166
malignant, viii, 26, 39, 41, 169, 173
mammogram, 134
management, x, xi, 13, 15, 71, 76, 95, 97, 105, 113
manipulation, 22, 114, 164, 166
MAPK, 24, 29, 30, 36, 50, 115
market, 116
Markov, 61, 66
Markov model, 61, 66
marrow, 48
Marx, 91
masking, 166
mastectomy, 133, 134
matrix, 78
maturation, 53
measurement, xii, 9, 141, 163, 164, 168
measures, 149
media, 90

median, 42, 57, 68, 119, 120, 122, 125, 126, 128, 129, 138, 139
mediators, ix, 75, 76, 77, 81
medication, 14, 86, 149, 150, 158
medications, 12, 76, 104
Medline, 149
men, 12, 13, 17
menopause, 7, 23, 78, 93, 115
menorrhagia, x, 76, 80, 81, 82, 83, 84, 85, 89
menstrual cycle, ix, 39, 44, 75, 78, 79, 81, 82, 84, 90, 91, 92, 93
menstruation, ix, 75, 76, 77, 78, 81, 82, 86, 89, 91
mesothelium, 109
messenger ribonucleic acid, 107
messenger RNA, 83, 172
meta-analysis, 57, 59, 72, 128, 143, 144, 161
metabolism, 9, 43
metabolites, 42, 43, 49, 88
metabolizing, 100
metastases, 126, 128
metastasis, 12, 14, 24
metastatic, viii, 3, 6, 8, 11, 15, 18, 21, 22, 27, 30, 32, 48, 62, 101, 122, 129, 138, 142, 144, 146, 172
metastatic cancer, 22
metastatic disease, 6, 8, 11, 22, 62, 101
metformin, 162
Mexico, 44
mice, 3, 13, 49, 53
microarray, 25, 34, 166, 170
microarray technology, 25
microenvironment, 115
mineralocorticoid, 116
minority, 7
miscarriage, 43, 91, 150
miscarriages, 149, 158
mitochondrial, 50
mitogen, 24, 29
mitogen-activated protein kinase, 24, 29
MMPs, 29
modalities, 97, 104
modality, xi, 114, 147
model system, 24, 26
modeling, ix, 55, 60, 68, 71
models, xii, 3, 13, 29, 30, 33, 60, 61, 62, 63, 65, 66, 138, 163
modulation, 22, 24, 32, 35, 90, 140
molecular biology, 82
molecular changes, 25
molecular markers, 135
molecular structure, 117
molecules, 24, 100
money, ix, 55, 56, 67, 70
monoclonal, 8, 30, 138
monoclonal antibodies, 30
monoclonal antibody, 8, 138
monotherapy, 30, 58, 68, 69, 72, 73, 125, 126, 127, 137, 138
mood, 101
mood change, 101
morbidity, 19, 123
mortality, 4, 19, 58, 59, 123, 135, 173
mother cell, 29
mothers, 46
mouse, 53
mRNA, 25, 79, 85, 93, 100, 168
mucus, 45, 47
muscle, viii, 7, 39, 40, 41, 47, 48, 100, 148
muscles, 2, 5, 24
musculoskeletal, 125, 126, 128, 129
mushrooms, 10
mutation, 3, 13, 28, 32
mutations, 3, 25
myalgia, 43, 101, 129
myocardial infarction, 10, 130, 131
myometrium, 84, 92

N

National Institutes of Health, 173
natural, vii, 2, 11, 24, 25, 78
natural killer, 78
natural killer cell, 78
nausea, 43, 101
neck, 10
needle aspiration, 38
neoplasm, 85, 88
neoplasms, 100, 173
neoplastic, 19
Netherlands, 44
New England, 144
New York, iii, iv
next generation, 91, 105
nicotinamide, 98
nitrogen, 40
NK cells, ix, 75, 174
NMR, 18
nodes, 42

normal, ix, xi, 9, 12, 25, 26, 40, 45, 48, 52, 53, 75, 81, 82, 86, 88, 104, 107, 110, 114, 141, 148, 163, 166
normalization, 87
North America, 37, 57, 92, 119, 120, 142
nuclear, 2, 12, 18, 50, 82, 83, 89, 114, 166
Nuclear factor, 89
nuclear magnetic resonance, 18
nuclear receptors, 2
nuclei, ix, 75, 82, 83, 89
nucleus, 24, 165
nutrition, 19

O

observations, 43, 67, 77, 97, 100
obstruction, 103, 106
OCs, 102, 104
oestrogen, xi, 16, 19, 23, 24, 33, 34, 37, 50, 113, 114, 115, 116, 119, 120, 122, 123, 125, 126, 127, 132, 133, 134, 135, 136, 137, 138, 139, 140, 141, 144, 145, 146, 172
Oncogene, 35, 50
oncology, 48, 71
Oncology, 8, 17, 38, 110, 113, 140, 142, 143, 144, 145, 146, 171, 173
online, 51
oocyte, 43, 52
oocytes, 53
oophorectomy, 97, 114
oral, viii, x, 3, 4, 15, 16, 19, 33, 39, 41, 42, 43, 49, 76, 82, 83, 85, 87, 90, 91, 92, 97, 104, 105, 108, 116, 123, 130, 138, 142
oral contraceptives, x, 76, 82, 83, 85, 90, 91, 92, 105
osteoblasts, 26
osteoclasts, 78
osteopenia, 3, 101
osteoporosis, 3, 8, 11, 17, 43, 101, 123, 126, 129, 135
ovarian cancer, vii, xi, 96, 163, 164, 165, 166, 167, 168, 169, 170, 171, 172, 173, 174
ovarian cancers, xi, 163, 164, 165, 168, 170, 171
ovarian cysts, 4, 5, 103, 104
ovarian failure, 133
ovarian follicular, ix, 39, 44
ovarian tumor, 173
ovariectomized, 35
ovariectomized rat, 35
ovaries, vii, x, xi, 1, 2, 11, 12, 24, 95, 96, 101, 114, 148, 149, 159, 163
ovary, viii, 24, 39, 40, 41, 44, 47, 50, 51, 52, 87, 98, 99, 108, 109, 115, 149, 159, 160, 161, 162, 166
overproduction, 96
oviduct, 46
ovulation, vii, viii, ix, xi, 39, 40, 42, 43, 44, 45, 46, 47, 51, 52, 53, 54, 86, 87, 133, 147, 148, 149, 150, 158, 159, 160, 161, 162
oxygen, 89

P

pain, x, 41, 81, 83, 84, 86, 87, 90, 95, 96, 97, 101, 102, 104, 105, 107, 108, 109, 130, 138
pain score, 104
palliative, 32
palpation, 134
paracrine, 98
parathyroid, 8
parathyroid hormone, 8
parenchyma, 24
Parkinson, 174
pathogenesis, x, 77, 81, 84, 89, 96, 108
pathogenic, 85
pathology, ix, 75, 76, 81, 82, 84, 86, 88, 89, 138
pathophysiology, 90, 91, 110
pathways, viii, 3, 5, 21, 24, 28, 29, 30, 32, 34, 36, 37, 41, 42, 138, 170
patient recruitment, 138
pelvic, x, 77, 78, 86, 90, 93, 95, 96, 102, 104, 105, 107, 108
pelvic pain, x, 90, 95, 96, 102, 104, 105, 107, 108
pelvis, 78, 79
peritoneal, ix, 75, 78, 79, 80, 90, 109
peritoneal cavity, ix, 75, 78
peritoneum, x, 78, 79, 95, 96
Peru, 44
phagocytosis, ix, 75, 78
pharmacodynamics, 17, 144
pharmacokinetic, 15
pharmacokinetics, 15, 71, 107, 144
pharmacological, 22, 23, 36, 106, 114
pharmacology, 16, 71, 107, 120
phenotype, 3, 26, 34
phosphate, 98
phosphorylates, 24
phosphorylation, 24, 33
physiological, 3, 44, 114

phytochemicals, viii, 2, 10, 19
PI3K, 24, 29, 30
PII, 26, 108
pilot study, 52, 104, 105, 107, 160, 161, 162
pituitary, ix, 5, 39, 41, 43, 44, 103, 114, 148
placebo, 4, 8, 33, 36, 37, 42, 43, 51, 60, 86, 97, 104, 119, 129, 135, 144
placenta, vii, 1, 40, 47
placental, 13, 98
planning, 110, 127
plants, viii, 2
plasma, 2, 6, 16, 23, 34, 36, 37, 47, 116, 139, 141, 142, 168, 174
plasma levels, 168
plasma membrane, 37
plasminogen, 170
plasticity, 32
platelet, 14
platelet count, 14
platinum, 164
plausibility, 57
play, 29, 76, 77, 78, 81, 85, 133, 138
polycystic ovarian syndrome, 51, 159, 160
polycystic ovary syndrome, 51, 52, 149, 159, 160, 161
polymerase, 48
polymerase chain reaction, 48
polymorphisms, 168, 173
polyp, 88, 93
polypeptides, 98
polyps, 81, 82, 86, 88, 89, 91, 93
poor, 43, 44, 51, 93, 138, 161
population, ix, 40, 46, 120, 123, 125, 139
population group, 139
positive correlation, 79
positive feedback, 81
powder, 79
power, 68
preclinical, 12, 30, 31
predictors, 13, 35
preference, 12
pregnancy, xi, 43, 44, 45, 46, 86, 87, 92, 93, 133, 147, 148, 149, 150, 158, 159, 162
pregnant, 46, 87
pregnant women, 46
premenopausal, xi, 5, 11, 22, 40, 96, 101, 102, 104, 105, 107, 109, 110, 113, 115, 130, 131, 132, 133, 140, 144, 145
premenopausal women, xi, 5, 11, 22, 40, 96, 101, 102, 104, 105, 109, 110, 113, 115, 130, 131, 132, 133, 140, 144, 145
press, 160
prevention, ix, 4, 8, 10, 11, 13, 14, 19, 40, 46, 108, 145, 146
primary tumor, 25, 170
primate, 52, 162
primates, 81
probability, 169
production, vii, viii, ix, x, 3, 5, 6, 7, 12, 24, 26, 37, 39, 40, 41, 43, 44, 45, 46, 51, 75, 76, 78, 79, 80, 81, 82, 84, 85, 89, 90, 91, 96, 97, 99, 100, 103, 106, 110, 115, 148
profit, 31
progesterone, x, 5, 12, 42, 50, 56, 75, 76, 82, 83, 85, 89, 97, 99, 106, 107, 110, 119, 126, 127, 135, 136, 137, 140, 146, 168, 170, 173
progestins, x, 22, 75, 83, 85, 86, 87, 89, 102, 104, 148
prognosis, 25, 138, 172, 173
program, 143
proinflammatory, x, 75
proliferation, 19, 24, 29, 30, 34, 76, 90, 102, 108, 114, 115, 139
promoter, 26, 50, 76, 78, 92, 99, 108, 109, 111
propagation, 29
property, iv, 123
prophylactic, 7
prophylaxis, 145
prostaglandin, 76, 77, 81, 82, 84, 89, 91, 110
prostaglandins, ix, 75, 76, 77, 78, 81, 86, 100
protection, 8, 19
protein, 5, 8, 24, 25, 26, 29, 34, 82, 83, 98, 99, 100, 107, 110, 114, 115, 164, 168, 170, 172
protein structure, 25
proteins, 24, 25, 29, 76, 110, 114, 170, 172
protocol, 9, 87, 126, 146, 160, 161, 162
puberty, 3
public, 159
public health, 159
pulmonary embolism, 122

Q

QALYs, 63, 69
QOL, 12, 103
quality of life, 68, 69, 70, 128, 129
quinones, 50

R

radiological, 169
radiotherapy, 114, 172
random, 156
range, 2, 7, 10, 23, 32, 62, 66, 77, 78
rapamycin, 30
rapamycin analogues, 30
rash, 101, 116, 117
rat, 35, 47, 48, 49
reaction mechanism, viii, 2
reality, 61
receptor-positive, viii, 13, 17, 21, 33, 34, 35, 36, 37, 38, 61, 71, 72, 110, 129, 143, 145, 146, 172
receptors, vii, viii, xi, 2, 3, 12, 13, 14, 16, 21, 24, 29, 35, 37, 44, 45, 50, 53, 90, 99, 106, 114, 140, 141, 148, 163, 164, 173
recognition, 37
recruiting, 139
recurrence, vii, ix, x, 4, 8, 11, 14, 15, 42, 55, 57, 58, 59, 60, 62, 63, 64, 65, 66, 67, 70, 73, 76, 77, 78, 92, 95, 97, 122, 125, 127, 128, 129, 130, 133, 134, 135, 137, 140, 142, 146
red wine, 10, 19
refractory, 27, 146, 174
regional, 57, 62, 66
regression, 79, 114, 116
regrowth, 164
regular, vii, 2, 7, 8
regulation, 24, 26, 32, 33, 35, 52, 53, 76, 92, 108, 110, 146, 148, 162, 167, 168, 172
regulators, 23
relapse, 8, 22, 29, 36, 64, 126, 132, 139
relapses, 61
relationship, viii, x, 2, 6, 14, 76, 79, 91
relevance, 10, 31, 89, 106, 158
remodeling, 8
repression, 24
repressor, 24
reproduction, 2, 47
reproductive age, x, 41, 95, 96, 103
resection, 88, 92
resistance, viii, 4, 13, 21, 24, 26, 27, 28, 30, 31, 32, 33, 34, 35, 36, 37, 39, 43, 120, 138, 146, 164
resistence, 28
resolution, 83
resource allocation, 71
responsiveness, 29, 90, 172
reticulum, 98
rings, 87
risk, viii, ix, 4, 7, 8, 10, 11, 17, 39, 42, 46, 49, 50, 55, 58, 59, 60, 62, 63, 64, 65, 66, 67, 70, 77, 85, 86, 88, 89, 93, 101, 106, 114, 122, 123, 127, 128, 129, 130, 133, 135, 138, 139, 140, 166, 168, 173
risk factors, 4, 130
risks, 7, 8, 12, 45, 60, 62, 63, 66, 67, 70, 129
robustness, 60, 61
rodents, 3
Rutherford, 172
rye, 10

S

S phase, 30
safety, vii, ix, 2, 7, 10, 12, 18, 33, 40, 41, 54, 72, 102, 105, 106, 129, 130, 143, 144
sample, xi, 67, 68, 147, 149, 151, 152, 153, 154, 155
Schiff, 28, 29, 30, 37
scores, 104
search, 71, 149
searches, 149
searching, 149
secretion, ix, 5, 15, 39, 41, 44, 99, 130, 148
sedative, 5
seed, 19
seeds, 10, 19
selective estrogen receptor modulator, viii, 21, 97
selectivity, vii, 1, 5, 41
sensitivity, 25, 28, 37, 44, 61, 62, 64, 65, 66, 69, 101, 166
septum, x, 87, 95, 96
sequencing, 11, 32, 72, 126, 127, 128, 143
series, 134, 140, 164, 165, 170
SERMs, 23, 115
serum, xii, 7, 17, 18, 35, 45, 49, 53, 104, 163, 169
services, iv
severity, ix, 75, 79, 91
sex, 49, 88, 100
sex hormones, 49
sex steroid, 100
sexuality, 93
shares, 100
shock, 24
short-term, 4, 62, 101
Short-term, 17

side effects, x, xi, 4, 8, 9, 11, 13, 87, 95, 97, 100, 101, 105, 113, 116, 120, 122, 125, 126, 128, 129, 133, 138, 140, 150, 158
signal transduction, 3, 30, 35, 36
signaling, 29, 30, 32, 33, 37, 42, 50, 164
signaling pathway, 32, 42
signaling pathways, 32, 42
signalling, 29, 50, 146
single nucleotide polymorphism, 168
single-nucleotide polymorphism, 173
sites, 24, 98, 99, 123
skeletal muscle, 48
skin, 5, 7, 40, 43, 98, 99, 101, 116, 117
soy, 10
Spain, 21
specificity, 6
spectroscopy, 18
spectrum, 66
spine, 8
stability, 169
stabilize, 25
stages, 52
statistics, 171
steroid, viii, 5, 9, 39, 40, 41, 47, 50, 53, 98, 99, 100, 109, 116, 168, 173
Steroid, 13, 15, 16, 33, 35, 36, 38, 47, 48, 49, 90, 106, 141, 146, 172, 173, 174
steroid hormone, 50, 99
steroidogenesis, 5, 9, 82
steroidogenic, 93, 98
steroids, x, 12, 41, 47, 89, 95, 96, 98
STI, 31
stimulus, 22, 81
strategies, viii, ix, x, xi, 12, 21, 23, 28, 30, 31, 37, 42, 55, 56, 57, 59, 60, 61, 62, 63, 64, 65, 66, 67, 68, 69, 70, 72, 73, 95, 96, 146, 163, 170
stroke, 4, 14, 122
stroma, 80, 81, 82, 90, 173
stromal, 6, 12, 85, 89, 90, 93, 99, 108, 109, 110, 141
stromal cells, 6, 89, 90, 93, 99, 108, 109, 110, 141
stromal fibroblasts, 85
subgroups, 11, 31, 58, 60, 67, 136, 170
substitution, 171
substrates, 98
success rate, 45
suffering, x, 95, 96
suicidal, 100
sulfate, 37
sulphate, 168
superiority, xi, 7, 22, 31, 46, 125, 147
suppression, xi, 6, 7, 15, 16, 76, 82, 83, 85, 105, 113, 116, 131, 132, 133, 140, 145, 148
suppressor, x, 19, 96
suppressors, 100
surgeons, 141
surgery, 27, 56, 59, 90, 104, 114, 122, 133, 134, 164, 172
surgical, x, 5, 28, 77, 78, 95, 96, 97, 104
survivability, 30
survival, 3, 4, 11, 14, 22, 27, 29, 30, 32, 38, 42, 53, 57, 58, 62, 64, 73, 101, 118, 119, 120, 122, 125, 126, 127, 128, 129, 132, 133, 134, 136, 138, 139, 142, 148, 173
survival rate, 32
switching, 11, 61, 68, 72, 126, 127, 128, 130, 132, 143, 144
symptoms, 4, 77, 80, 81, 82, 83, 84, 85, 89, 97, 102, 104, 109, 130, 168
syndrome, ix, 40, 46, 51, 52, 149, 159, 160, 161
synergistic, 30
synthesis, vii, viii, 1, 2, 3, 9, 18, 22, 36, 39, 41, 43, 46, 48, 78, 81, 82, 85, 98, 100, 102, 115, 116, 132
systemic circulation, 41

T

Taiwan, 44
tamoxifen, vii, viii, ix, xi, 2, 4, 5, 6, 7, 8, 9, 10, 11, 13, 14, 15, 16, 17, 18, 19, 21, 22, 23, 27, 28, 29, 30, 31, 32, 33, 34, 35, 36, 37, 38, 41, 42, 48, 51, 55, 56, 71, 72, 73, 101, 102, 105, 107, 108, 113, 114, 115, 118, 119, 120, 121, 122, 123, 124, 125, 126, 127, 128, 129, 130, 132, 133, 134, 135, 136, 137, 138, 139, 142, 143, 144, 145, 146, 148, 164, 167, 169, 170, 171, 172
targets, 30
taxane, 164
taxonomy, 32
testes, 2
testosterone, x, 40, 48, 88, 96, 98, 115, 148
Testosterone, 93
Thailand, 44
therapeutic agents, 5
therapeutic interventions, 8
therapeutics, 105

threatening, 4, 7, 122
threshold, 57, 65, 66, 67, 71
thresholds, 57, 61, 65, 71
thromboembolic, 4, 6, 7, 120, 122, 123, 125, 126, 129, 133
thromboembolism, 64
thrombosis, 4, 14
thrombotic, 171
thyroid, 41
TIA, 126
tibolone, 88, 89, 93
time frame, 62, 63
timing, 116
tissue, vii, viii, x, 1, 2, 3, 5, 12, 24, 26, 33, 39, 40, 41, 43, 47, 48, 49, 77, 79, 81, 82, 86, 88, 92, 95, 96, 97, 98, 99, 100, 106, 107, 109, 115, 137, 141, 148, 164, 165, 166, 168, 174
TMA, 166
TNF, 170
Tokyo, 1
total cholesterol, 9
toxic, 4, 40, 164
toxic effect, 4, 40
toxicities, 40, 60
toxicity, 5, 22, 28, 32, 42, 138, 139
trans, 11, 34
transcript, 50, 170
transcription, x, 2, 24, 25, 48, 50, 76, 83, 86, 99, 108, 111
transcription factor, 24, 99, 111
transcription factors, 24, 99
transcriptional, 33, 35, 82, 89
transcripts, 48, 79
transfer, 5, 91, 93
transformation, 26, 76
transition, 30, 64, 69
translation, 34
translational, 48
translocation, 82, 89
transmembrane, 138
transmembrane glycoprotein, 138
trastuzumab, 30, 32, 138, 146
triglyceride, 9
tumor, vii, xii, 5, 6, 10, 14, 16, 23, 25, 26, 27, 28, 30, 32, 33, 34, 38, 42, 48, 50, 58, 70, 85, 102, 113, 114, 115, 121, 130, 135, 137, 139, 141, 163, 164, 165, 166, 168, 171, 173
tumor cells, 5, 6
tumor growth, 10
tumor resistance, 32
tumors, 4, 5, 6, 10, 13, 25, 28, 29, 32, 42, 47, 49, 85, 164, 169, 170, 171, 173
turnover, 7, 34
two-way, 61, 62, 65, 66
tyrosine, 24, 29, 30, 31, 50, 138

U

ubiquitin, 170
ultrasound, 134
uncertainty, 61, 62, 67
United Kingdom, 163
United States, 4, 14, 100, 114, 135
urokinase, 170
uterine cancer, 4, 10
uterine fibroids, 92
uterus, 35, 81, 96

V

validation, 73
values, 9, 27, 150, 170
variables, 60
variation, 168
vascular endothelial growth factor, 30, 90
vascular endothelial growth factor (VEGF), 30
vascularization, 85
vasomotor, 4
vegetables, 10
VEGF, x, 30, 76, 85
velocity, 85
Victoria, 51
vimentin, 170
vitamin D, 102

W

wealth, 11
weight gain, 9
WHO, 43
wine, 10, 19
withdrawal, 5, 22, 29, 76, 82, 89, 120, 130
workers, 29
WT1, 108

X

xenograft, 10, 172

Z

Zoledronic acid, 143